T0215036

Guide to Modern Physics

This is a "how to guide" for making beginning calculations in modern physics. The academic level is second year college physical science and engineering students. The calculations are performed in Mathematica, and stress graphical visualization, units, and numerical answers. The techniques show the student how to learn the physics without being hung up on the math. There is a continuing movement to introduce more advanced computational methods into lower-level physics courses. Mathematica is a unique tool in that code is written as "human readable" much like one writes a traditional equation on the board.

Key Features:
- Concise summary of the physics concepts.
- Over 300 worked examples in Mathematica with available notebooks.
- Tutorial to allow a beginner to produce fast results.

James Rohlf is a Professor at Boston University. As a graduate student, he worked on the first experiment to trigger on hadron jets with a calorimeter, Fermilab E260. His thesis (G. C. Fox, advisor, C. Barnes, R. P. Feynman, R. Gomez) used the model of Field and Feynman to compare observed jets from hadron collisions to that from electron-positron collisions and made detailed acceptance corrections to arrive at first the measurement of quark-quark scattering cross sections. His thesis is published in *Nuclear Physics* B171 (1980) 1. At the Cornell Electron Storage Rings, he worked on the discovery of the Upsilon (4S) resonance and using novel event shape variables developed by Steven Wolfram and his thesis advisor, Geoffrey Fox. He performed particle identification of kaons and charmed mesons to establish the quark decay sequence, b –> c. At CERN, he worked on the discovery of the W and Z bosons and measurement of their properties. Presently, he is working on the Compact Muon Solenoid (CMS) experiment at the CERN Large Hadron Collider (LHC) which discovered the Higgs boson and is searching for new phenomena beyond the standard model.

Guide to Modern Physics
Using Mathematica for Calculations and Visualizations

James W. Rohlf

CRC Press
Taylor & Francis Group
Boca Raton London New York

CRC Press is an imprint of the
Taylor & Francis Group, an **informa** business

Designed cover image: James W. Rohlf

First edition published 2024
by CRC Press
2385 NW Executive Center Drive, Suite 320, Boca Raton FL 33431

and by CRC Press
4 Park Square, Milton Park, Abingdon, Oxon, OX14 4RN

CRC Press is an imprint of Taylor & Francis Group, LLC

ISBN: 978-1-032-49801-0 (hbk)
ISBN: 978-1-032-49686-3 (pbk)
ISBN: 978-1-003-39551-5 (ebk)

DOI: 10.1201/9781003395515

Typeset in Nimbus Roman
by KnowledgeWorks Global Ltd.

Publisher's note: This book has been prepared from camera-ready copy provided by the authors.

Contents

Preface

Modern physics uses combinations of physical constants which are unwieldy in both magnitude and units. The beginning student can easily be over-whelmed in the details of calculations that obscure the underlying physics at precisely the time when getting a numerical answer is the key to learning. This book is a guide to how to perform these calculations in Mathematica and includes many examples that will be encountered in a first course in modern physics. Mathematica is used as a data base for physical constants, a calcula-tor with units, an algebraic manipulator with math functions, an integrator, a differentiator, an equation solver, a series expander, a plotter, and more.

Appendix A is intended to help a beginner immediately get started in Mathematica. One of the most important things to know at the outset is how physical constants are stored and evaluated. The Mathematica names, sym-bols, and numerical values of important quantities are given in App. B.

The body of the text puts the calculations in physics context. The cal-culations are displayed as numbered examples, followed by executable code labeled as In[ ]:=. The output is labeled Out[ ].

Example 5.3 Calculate the numerical value of the Bohr radius ($n = 1$).

In[3]:= m = | **electron** PARTICLE | | *mass* | ;

$$N\left[\text{UnitConvert}\left[\frac{4\,n^2\,\pi\,\hbar^2\,\varepsilon_\theta}{e^2\,m}, \text{nm}\right] \,/.\, n \rightarrow 1, 3\right]$$

Out[4]= 0.0529 nm

The beauty of Mathematica is that the code is human readable because the core part of the calculation *looks* like what you would see printed in a book. No calculation should ever be done in physics without checking the units. Mathematica has the extremely useful built in property that code will not execute if the units do not match. In the example above, the expression,

$$\frac{4n^2\pi\hbar^2\epsilon_0}{e^2m},$$

will not evaluate if its units are incompatible with the requested unit of nanometers (nm). If the unit is not specified, the output will be in the international system of units (SI), no matter what units were assigned to the input. Thus, Mathematica is a powerful unit converter. This can be a huge time saver.

The book contains a number of figures that are all drawn in Mathematica.

The chapters are largely independent of one another, except that the material in Chap. 1 should be digested first, and Chap. 6 (Particle in a Box) should come before Chaps. 7 (Quantum Harmonic Oscillator) or 8 (Hydrogen Atom).

CHAPTER 1

Basis of Modern Physics

1.1 CHARGE AND THE ELECTRONVOLT

Charge is an intrinsic property of the electron. There is no such thing as an electron without its charge. No free particles have ever been observed with fractional electron charge.

1.1.1 Elementary Charge

The function Quantity["ElementaryCharge"] is used to acquire the fundamental electric charge in Mathematica. The charge is elegantly displayed as e and has units assigned. The quantity e is positive, so the proton charge is e and the electron charge is $-e$. Details for how to get a physical constant with units using the alternate natural language box are given in A.5 and a list of constants is given in App. B.

Example 1.1 Get the elementary charge.

In[1]:= **Quantity["ElementaryCharge"]**

Out[1]= *e*

The output of Ex. 1.1 is written in a slightly different shade e to distinguish its representation of a unit from an italic e.

The unit of charge is the coulomb (C) which is an amp (A) times a second (s). The coulomb unit is acquired with the function Quantity["Coulombs"].

DOI: 10.1201/9781003395515-1

Example 1.2 Get the unit coulomb.

In[2]:= **Quantity["Coulombs"]**

Out[2]= **1 C**

The function UnitConvert[ ] converts the symbolic unit to a number in a specified unit. The function N[ ] displays that number in decimal form to the specified number of digits. It is important to note that the first argument of UnitConvert[e, C] displayed in the input cell of ex. 1.3, is a stored unit as generated in ex. 1.1. It is the abbreviation for Quantity["ElementaryCharge"]. The second argument is the coulomb unit obtained in ex. 1.2. It is the abbreviation for Quantity["Coulombs"].

Example 1.3 Get the numerical value of the elementary charge in C to 4 figures.

In[3]:= **N[UnitConvert[e, C], 4]**

Out[3]= **1.602 × 10^{-19} C**

The preferred unit of energy in modern physics is the electronvolt (eV), defined as the kinetic energy gained by an electron when accelerated through a potential difference of 1 volt (V). The eV is the energy scale of outer electrons in atoms. Similarly, keV is the scale of inner electrons in atoms, MeV is the scale of protons in the nucleus, and GeV is the scale of energetic quarks in the proton. There is a subtle and significantly important property of this unit in that the elementary charge (e) × 1 V is equal to 1 eV. One may take advantage of this in many numerical calculations.

For the input cell of ex. 1.4, V is the unit from Quantity["Volts"] and eV is the unit from Quantity["ElectronVolts"].

Example 1.4 Compare e times 1 V to 1 eV (the "==" perfoms a logical comparison).

In[4]:= **1 e V == 1 eV**

Out[4]= **True**

One e per coulomb (C) is equal to 1 eV per joule (J). Thus, the elementary charge of 1.602×10^{-19} C gives the eV to be 1.602×10^{-19} J.

Example 1.5 Compare the units $\frac{e}{C}$ to $\frac{eV}{J}$ and calculate the numerical value of the ratio.

In[5]:= $\dfrac{e}{C} == \dfrac{eV}{J}$

$$\texttt{NumberForm}\left[\texttt{N}\left[\frac{e}{C}, 4\right]\right]$$

Out[5]= **True**

Out[6]//NumberForm=
        $\mathbf{1.602 \times 10^{-19}}$

1.1.2 Strength of Electromagnetism

The strength of the electric force between 2 elementary charges (e) is written

$$F = \frac{e^2}{4\pi\varepsilon_0 r^2},$$

where r is the separation distance and ε_0 is the electric constant.

Example 1.6 Get the electric force constant.

In[7]:= **Quantity["ElectricConstant"]**

Out[7]= ε_0

The preferred unit of $\frac{e^2}{4\pi\varepsilon_0}$ in modern physics is eV·nm.

Example 1.7 Calculate the value of $\frac{e^2}{4\pi\varepsilon_0}$ in units eV·nm.

In[8]:= $\texttt{N}\left[\texttt{UnitConvert}\left[\dfrac{e^2}{4\pi\varepsilon_0}, \texttt{eV nm}\right], 3\right]$

Out[8]= **1.44 nm eV**

The units are always output in the same order (mass, length, time, *etc.*) no matter what order they are input.

The unit eV/nm is a unit of force that is practical to use at the atomic scale. The attractive force between an electron and proton separated by a distance of 0.1 nm is

$$F = \frac{e^2}{4\pi\varepsilon_0 r^2} = \frac{1.44\text{eV}\cdot\text{nm}}{(0.1\ \text{nm})^2} = \frac{14.4\ \text{eV}}{0.1\ \text{nm}}.$$

The strength of the electric force in these units can be identified with the atomic energy and distance scales.

1.2 PLANCK'S CONSTANT

At the heart of modern physics lies the fact that measured quantities such as energy and angular momentum are not continuous. Instead, they take on discrete values, a concept referred to as quantization. Planck's constant (h) sets the scale for this granularity which is not observable in classical physics. The unit of Planck's constant is that of momentum × distance or energy × time,

$$\left(\frac{\text{kg} \cdot \text{m}}{\text{s}}\right)(\text{m}) = (\text{J} \cdot \text{s}).$$

Planck's constant is extremely tiny compared to $1\ \text{kg} \cdot \text{m}^2/\text{s}$. Even compared to a microscopic scale, $1\ \mu\text{g} \cdot (\mu\text{m})^2/\text{s}$, Planck's constant is still very tiny which is why the quantization goes unnoticed in the classical regime.

Example 1.8 Get Planck's constant.

In[9]:= **Quantity["PlanckConstant"]**

Out[9]= h

The function NumberForm[ ] is one of several ways to output a numerical value.

Example 1.9 Get the numerical value of Planck's constant in SI units.

In[10]:= **NumberForm[N[UnitConvert[h], 3]]**

Out[10]//NumberForm=
$6.63 \times 10^{-34}\ \text{kg}\,\text{m}^2/\text{s}$

A more suitable unit is eV·s.

Example 1.10 Get the numerical value of Planck's constant in $\text{eV} \cdot \text{s}$.

In[11]:= **NumberForm[N[UnitConvert[h, eV s], 3]]**

Out[11]//NumberForm=
$4.14 \times 10^{-15}\ \text{s}\,\text{eV}$

1.2.1 The Combination hc

Planck's constant appears naturally together with the speed of light (c).

Example 1.11 Get the speed of light.

```
In[12]:= Quantity["SpeedOfLight"]
Out[12]=
        c
```

Example 1.12 Get the numerical value of the speed of light in vacuum.

```
In[13]:= UnitConvert[ c ]
Out[13]=
        299 792 458 m/s
```

The product hc is relatively easy to remember and is extremely useful in calculations involving h.

Example 1.13 Get Planck's constant times the speed of light.

```
In[14]:= N[UnitConvert[ h c , eV nm ], 4]
Out[14]=
        1240. nm eV
```

For higher energies of the nuclear scale, one may note that

$$1240 \text{ eV} \cdot \text{nm} = 1240 \text{ MeV} \cdot \text{fm}.$$

The numerical value of hc is revealing that smaller distances are associated with higher energies. This is the central theme of modern physics.

1.2.2 Reduced Planck's Constant ($\hbar$)

The quantization is commonly written in terms of the "reduced" Planck's constant ($\hbar$), called "h-bar",

$$\hbar = \frac{h}{2\pi}.$$

Example 1.14 Get the reduced Planck's constant.

```
In[15]:= Quantity["ReducedPlanckConstant"]
Out[15]=
        ℏ
```

The quantity $\hbar c$ appears frequently and is also easy to remember.

Example 1.15 Get the reduced Planck's constant times the speed of light.

In[16]:= `N[UnitConvert[ ℏ c , eV nm ], 3]`

Out[16]=

  197. nm eV

Again,

$$197\,\text{eV}\cdot\text{nm} = 197\,\text{MeV}\cdot\text{fm}.$$

1.2.3 Fine Structure Constant

The strength of the electromagnetic force is made dimensionless by dividing by $\hbar c$, giving the famous fine structure constant (α),

$$\alpha = \frac{\left(\frac{e^2}{4\pi\varepsilon_0}\right)}{\hbar c}.$$

Example 1.16 Get the fine structure constant.

In[17]:= `Quantity["FineStructureConstant"]`

Out[17]=

  α

The fine structure constant is historically written as the ratio of integers. The function Rationalize [ ] will do this.

Example 1.17 Calculate α, rationalized to order 10^{-5}.

In[18]:= `Rationalize[UnitConvert[` $\dfrac{e^2}{4\,\pi\,\varepsilon_0\,\hbar\,c}$ `], 10`$^{-5}$`]`

Out[18]=

  $\dfrac{1}{137}$

The fine structure constant of Ex. 1.16 can be compared with $e^2/4\pi\varepsilon_0\hbar c$.

Example 1.18 Compare the quantity α to $\frac{e^2}{4\pi\varepsilon_0\hbar c}$.

In[19]:= α `==` $\dfrac{e^2}{4\,\pi\,\varepsilon_0\,\hbar\,c}$

Out[19]=

  True

The fine structure constant appears frequently in atomic physics which is described by the electromagnetic force strength combined with the quantization in terms of Planck's constant. For example, the root-mean-square (rms) speed v of an electron in the hydrogen atom is

$$v = \alpha c.$$

Example 1.19 Calculate the rms speed and kinetic energy of an electron in hydrogen.

In[20]:= m = [**electron** PARTICLE] [[*mass*]]; N[UnitConvert[v = α c], 3]

N[UnitConvert[$\frac{1}{2}$ m v^2, eV], 3]

Out[20]=

2.19 × 10^6 m/s

Out[21]=

13.6 eV

The atomic electron is not relativistic.

1.2.4 Strength of Gravity

The gravitational force between an electron and a proton can be written

$$F = \frac{Gm_e m_p}{r^2}$$

Example 1.20 Get the universal gravational constant G.

In[22]:= Quantity["GravitationalConstant"]

Out[22]=

G

The strength of gravity between and electron and proton can be made dimensionless with the expression $\frac{Gm_e m_p}{\hbar c}$ which gives the gravitational equivalent of α. The particle masses may be obtained with the natural language box (A.5). They are discussed further in 1.3.1.

Example 1.21 Calculate the dimensionless strength of gravity.

In[23]:= **m** = [**electron** PARTICLE] [*mass*] ;

M = [**proton** PARTICLE] [*mass*] ;

$$\text{NumberForm}\left[\text{UnitConvert}\left[\frac{G\ m\ M}{\hbar\ c}\right], 1\right] \cdot$$

Out[25]//NumberForm=

$$3. \times 10^{-42}$$

Gravity is 40 orders of magnitude weaker than the electric force.

1.3 ENERGY

1.3.1 Mass and Momentum Units

A convenient mass unit is MeV/c^2, where c is the speed of light in vacuum. Masses may be obtained with either the function Quantity[ ] or the natural language box.

Example 1.22 Get the electron mass with both the Quantity[ ] function and the natural language box.

In[26]:= **Quantity["ElectronMass"]**

[**electron** PARTICLE] [*mass*]

Out[26]=

$$m_e$$

Out[27]=

$$510.9989461\ keV/c^2$$

Example 1.23 Compare the 2 methods of acquiring the electron mass.

In[28]:= m_e == [**electron** PARTICLE] [*mass*]

Out[28]=

True

These representations of the mass are used interchangeably in this work.

Example 1.24 Get the proton mass with both the Quantity[ ] function and the natural language box.

In[29]:= `Quantity["ProtonMass"]`

$$\boxed{\textbf{proton} \ \text{PARTICLE}} \left[\boxed{\textit{mass}} \right]$$

Out[29]=

$$m_p$$

Out[30]=

938.2720813 MeV/c^2

The preferred momentum unit is eV/c (or keV/c, MeV/c, *etc.*) depending on the scale. This is convenient whether or not the electron is relativistic. The formula for relativistic momentum is given in 1.3.3.

Example 1.25 Calculate the momentum in eV/c for an electron with speed $v = 10^4$ m/s.

In[31]:= `v = 10`^{4}` m / s;`

$$N\left[\text{UnitConvert}\left[m_e \ v, \ \frac{eV}{c}\right], 3\right]$$

Out[32]= 17.0 eV/c

The relationship between momentum (p) and nonrelativistic kinetic energy (K) is

$$K = \frac{p^2}{2m}.$$

Example 1.26 Calculate the momentum in MeV/c of an electron whose kinetic energy is 10 keV.

In[33]:= `K = 10 keV ;`

$$N\left[\text{UnitConvert}\left[\sqrt{2 \ m_e \ K}, \ \frac{MeV}{c}\right], 2\right]$$

Out[34]=

0.10 MeV/c

1.3.2   Kinetic and Mass Energy

There are two forms of energy: energy due to motion (kinetic energy) and energy stored as mass (mass energy). The mass energy (E_0) is

$$E_0 = mc^2.$$

The total energy (E) is

$$E = mc^2 + K.$$

1.3.3 Momentum

If the particle is relativistic, the rule for momentum (as observed to be conserved in the lab) is

$$p = \frac{mv}{\sqrt{1 - \frac{v^2}{c^2}}}.$$

and the total energy is

$$E = \sqrt{(mc^2)^2 + (pc)^2} = mc^2 \sqrt{1 + \frac{p^2}{m^2 c^2}}.$$

The second term inside the square root is a small number if $v \ll c$, and the kinetic energy reduces to the familiar nonrelativistic form.

Example 1.27 Get the nonrelativistic kinetic energy from the relativistic energy equation.

```
In[35]:=  ClearAll["Global`*"];
          $Assumptions = {m > 0, c > 0};
          K = √((m c²)² + (p c)²) - m c²;
          Series[K, {p, 0, 2}] // Simplify
Out[36]=
          p²
          ──  + O[p]³
          2 m
```

In Ex. 1.27, the variables are user defined (not units).

Example 1.28 Calculate the fractional error in the nonrelativistic formula for momentum when $v = 0.1\ c$.

```
In[37]:=  p = ────────────── ;   ──────── /. v → 0.1 c
              √(1 - (v/c)²)        p - m v
                                      p
Out[37]=
          0.00501256
```

The expression is good to 0.5% when $v = 0.1\,c$, so that is a good approximate number for the boundary at which one can get a realistic answer without using the relativistic expression.

The exact expression for the kinetic energy is

$$K = \sqrt{(mc^2)^2 + (pc)^2} - mc^2$$

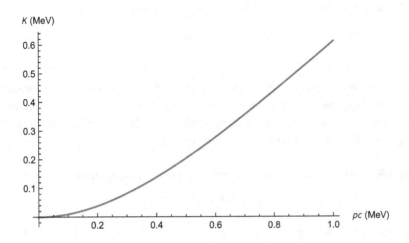

Figure 1.1 The plot shows the kinetic energy of an electron *vs.* momentum times *c*. At low momentum the expression is quadratic but at high momentum it is linear.

Example 1.29 Calculate the kinetic energy and speed of an electron with momentum 1 MeV/*c*.

In[38]:= m = [**electron** PARTICLE] [*mass*] ; p = 1 MeV/c ;

$$N\left[UnitConvert\left[\sqrt{(m\,c^2)^2 + (p\,c)^2} - m\,c^2,\ MeV\right], 3\right]$$

$$N\left[UnitConvert\left[\frac{p\,c}{\sqrt{(m\,c^2)^2 + (p\,c)^2}}\ c\right], 3\right]$$

Out[39]=

0.612 MeV

Out[40]=

2.67 × 10⁸ m/s

1.3.4 Potential Energy

There is another energy that is called the potential energy U. It is not a new type of energy but is a binding energy related to the mass and kinetic energies. In the description of the hydrogen atom, the mass energy is left out and the total energy is written as the sum of kinetic (Ex. 1.19) plus potential,

$$K + U = -13.6 \text{ eV},$$

where the potential energy has been calculated from

$$U = -\frac{e^2}{4\pi\varepsilon_0 r}.$$

(In the language of quantum mechanics, both the position r and the momentum p are smeared out and the above energies become averages.) The negative sign means that the force is attractive and the electron is bound. An energy of 13.6 eV must be added to free the electron.

Example 1.30 Get the energy needed to ionize a hydrogen atom.

In[41]:= [**hydrogen** ELEMENT] [*ionization energies*]

Out[41]=
$\{ 13.598 \text{ eV} \}$

This means that

$$m_{\text{H}}c^2 + 13.6 \text{ eV} = m_{\text{e}}c^2 + m_{\text{p}}c^2$$

and using eV/c^2 as a mass unit,

$$m_{\text{H}} = m_{\text{e}} + m_{\text{p}} - 13.6 \text{ eV}/c^2$$

The mass of the hydrogen atom is smaller than the sum of the masses of its parts. The value of U is negative.

$$U = m_{\text{H}}c^2 - (m_{\text{e}}c^2 + m_{\text{p}}c^2) - K$$

In the lowest energy (most bound) state of the hydrogen atom, the (average) electron kinetic energy is

$$K = 13.6 \text{ eV}$$

and the (average) potential energy is

$$U = -27.2 \text{ eV}$$

In the describing atom, one doesn't worry about the constant mass energies of the electron and proton and they are left out of the expression for total energy,

$$E = U + K,$$

This is understood to mean

$$U = -27.2 \text{ eV} + m_e c^2 + m_p c^2,$$

and

$$E = U + K = m_H c^2.$$

One has to know from the context (GeV vs. eV scale), if the shortcut of leaving out the mass part has been made.

1.3.5 Nuclear Binding Energy

The binding energy of an atomic nucleus is calculated by summing the mass energies of all the neutrons and protons and subtracting the mass energy of the nucleus.

Example 1.31 Get the neutron mass.

In[42]:= `Quantity["NeutronMass"]`

$$\boxed{\textbf{neutron}\ \text{PARTICLE}}\Big[\,\boxed{\textit{mass}}\,\Big]$$

Out[42]=

$$m_n$$

Out[43]=

$$939.5654133 \text{ MeV}/c^2$$

The neutron mass is larger than the proton mass (Ex. 1.24) by about 1.3 MeV/c^2.

The alpha particle is the nucleus of helium and is composed of 2 neutrons and 2 protons. It is very stable.

Example 1.32 Calculate the binding energy of the alpha particle.

In[44]:= `UnitConvert`$\Big[$

$$\Big(2\,\boxed{\textbf{neutron}\ \text{PARTICLE}}\Big[\,\boxed{\textit{mass}}\,\Big] + 2\,\boxed{\textbf{proton}\ \text{PARTICLE}}\Big[\,\boxed{\textit{mass}}\,\Big] -$$

$$m_\alpha\Big)\ c^2,$$

`MeV`$\Big]$

Out[44]= `28.29558 MeV`

The uranium-238 nucleus has 92 protons and 146 neutrons.

Example 1.33 Calculate the binding energy of ^{238}U.

In[45]:= **UnitConvert**[

$\Big($ 146 [**neutron** PARTICLE] [*mass*] + 92 [**proton** PARTICLE] [*mass*] -

$\big($ ([**uranium-238** ISOTOPE] [*atomic mass*])

- 92 [**electron** PARTICLE] [*mass*])) c^2, MeV]

Out[45]= **1801.69 MeV**

Notice that Mathematica reports the atomic mass of ^{238}U so the electrons need to be subtracted from the atomic mass to get the nuclear mass.

A neutron is added to an oxygen-16 nucleus to make an oxygen-17 nucleus,

$$n + {}^{16}O \rightarrow {}^{17}O + \text{energy.}$$

Example 1.34 Calculate the binding energy of the neutron.

In[46]:= **UnitConvert**[(([**neutron** PARTICLE] [*mass*] +

([**oxygen-16** ISOTOPE] [*atomic mass*]

- 16 [**electron** PARTICLE] [*mass*]) -

([**oxygen-17** ISOTOPE] [*atomic mass*]

- 17 [**electron** PARTICLE] [*mass*])) c^2, MeV]

Out[46]=

  4.6541 MeV

1.3.6 Q Value

The energy released in a decay (Q) is the mass energy of the decaying paricle minus the sum of the mass energies of the decay products. The neutron decays

$$n \rightarrow p + e + \bar{\nu}_e.$$

The neutrino ($\bar{\nu}_e$) is essentially massless. This process is called β^- decay.

Example 1.35 Calculate the Q value for neutron decay.

In[47]:= `UnitConvert[((`**neutron** PARTICLE`)[`*mass*`] -`
`(`**proton** PARTICLE`)[`*mass*`] +`
electron PARTICLE`)[`*mass*`])) c`^{2}`, MeV]`

Out[47]=
0.782333 MeV

The decay cannot occur if Q is negative, but it is possible for a bound proton to convert into a neutron inside the nucleus,

$$p \to n + e^+ + \nu_e.$$

This process is called β^+ decay. The positron (e^+) has the same mass as the electron. The β^+ process cannot happen for a free proton because the proton mass is smaller than the neutron mass. An example is the decay of boron-8,

$$^8B \to {}^8Be + e^+ + \nu_e.$$

Example 1.36 Calculate Q for β^+ decay of ^8B.

In[48]:= `UnitConvert[((`**boron-8** ISOTOPE`)[`*atomic mass*`] - 5 m`$_e$` -`
`(`**beryllium-8** ISOTOPE`)[`*atomic mass*`] - 4 m`$_e$` +`
electron PARTICLE`)[`*mass*`])) c`^{2}`, MeV]`

Out[48]=
16.96 MeV

Another example of the same process is the fusion of 2 protons to make a proton-neutron bound state called the deuteron (d),

$$p + p \to d + e^+ + \nu_e.$$

Example 1.37 Calculate Q for the fusion of 2 protons.

In[49]:= `UnitConvert[(2 `**proton** PARTICLE`)[`*mass*`] -`
`(`**deuteron** PARTICLE`)[`*mass*`] +`
electron PARTICLE`)[`*mass*`])) c`^{2}`, MeV]`

Out[49]=
0.420236 MeV

Note that Q of the process is subtly different from the binding energy of the deuteron because the positron mass must be included. The positron will annihilate with an electron and produce energy in the form of 2 photons.

The fusion process generally occurs for elements lighter than iron, but does not occur for heavier elements.

Example 1.38 Show that the decay $^{230}U \rightarrow ^{229}Pa + p + e + \bar{\nu}_e$ cannot occur.

In[50]:= UnitConvert[(((uranium-230 ISOTOPE) [atomic mass] - 92 m_e) -

((protactinium-229 ISOTOPE) [atomic mass] - 91 m_e

+ proton PARTICLE) [mass] +

electron PARTICLE) [mass])) c^2, MeV]

Out[50]=

-6.1 MeV

The Q value is negative so the decay cannot occur.

Example 1.39 Show that the decay $U^{230} \rightarrow U^{229} + n + e^+ + \nu_e$ cannot occur.

In[51]:= UnitConvert[(((uranium-230 ISOTOPE) [atomic mass] - 92 m_e) -

((uranium-229 ISOTOPE) [atomic mass] - 92 m_e

+ neutron PARTICLE) [mass] +

electron PARTICLE) [mass])) c^2, MeV]

Out[51]=

-8.2 MeV

The Q value is negative so the decay cannot occur.

1.4 THE PHOTON

1.4.1 Wavelength and Energy

A photon is a particle (quantum) of light. It has zero mass so its energy is pure kinetic. The photon energy as a function of wavelength (λ) is

$$E = \frac{hc}{\lambda}$$

Example 1.40 Calculate the energy range of visible photons (wavelengths 400 nm to 700 nm).

In[52]:= $\mathtt{N}\left[\left\{\mathtt{UnitConvert}\left[\dfrac{\mathtt{h\ c}}{\mathtt{700\ nm}},\ \mathtt{eV}\ \right],\ \mathtt{UnitConvert}\left[\dfrac{\mathtt{h\ c}}{\mathtt{400\ nm}},\ \mathtt{eV}\ \right]\right\},\ 3\right]$

Out[52]=

$\{\ 1.77\ \mathtt{eV}\ ,\ \ 3.10\ \mathtt{eV}\ \}$

The wavelength of a 10 keV x ray is

$$\lambda = \frac{hc}{E} = \frac{1240\ \text{eV}\cdot\text{nm}}{10^4\ \text{eV}} = 0.124\ \text{nm}$$

and a 1 MeV γ ray has

$$\lambda = \frac{1240\ \text{eV}\cdot\text{nm}}{10^6\ \text{eV}} = 1240\ \text{fm}$$

Example 1.41 Calculate the energy in meV of a 3 cm wavelength microwave photon.

In[53]:= $\mathtt{N}\left[\mathtt{UnitConvert}\left[\dfrac{\mathtt{h\ c}}{\mathtt{3\ cm}},\ \mathtt{meV}\ \right],\ 3\right]$

Out[53]=

$0.0413\ \mathtt{meV}$

1.4.2 Speed and Frequency

The relationship between wave speed (c), wavelength (λ), and frequency (f) is

$$c = \lambda f$$

Photon frequency is usually given in hertz (Hz) which is just an inverse second.

Example 1.42 Calculate the frequency range of visible photons.

In[54]:= $\mathtt{N}\left[\left\{\mathtt{UnitConvert}\left[\dfrac{\mathtt{c}}{\mathtt{700\ nm}},\ \mathtt{Hz}\ \right],\ \mathtt{UnitConvert}\left[\dfrac{\mathtt{c}}{\mathtt{400\ nm}},\ \mathtt{Hz}\ \right]\right\},\ 3\right]$

Out[54]=

$\left\{\ 4.28 \times 10^{14}\ \mathtt{Hz}\ ,\ \ 7.49 \times 10^{14}\ \mathtt{Hz}\ \right\}$

The relationship between energy and frequency is

$$E = hf$$

For example, the energy of photons from a 100 MHz radio broadcast is

$$E = hf = (4.14 \times 10^{-15} \text{ eV} \cdot \text{s})(100 \times 10^{6} \text{s}^{-1}) = 4.14 \times 10^{-7} \text{ eV}$$

1.5   DE BROGLIE WAVELENGTH

1.5.1   Particle Wave Duality

The light wave has a particle (photon) interpretation, and the electron has a wave behavior. *This is the single most important concept in modern physics.* It is the small mass of the electron that makes its wave properties observable. The de Broglie wavelength of any particle is defined in terms of its momentum as

$$\lambda = \frac{h}{p} = \frac{hc}{pc}.$$

The expression works for relativistic particles, provided one uses the relativistic expression for momentum. Note that for a photon, $E = pc$, resulting in the usual formula, $E = \frac{hc}{\lambda}$.

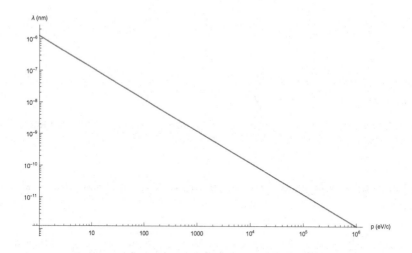

Figure 1.2   The wavelength is shown *vs.* momentum on a log-log plot.

Example 1.43 Calculate the de Broglie wavelength of a an electron whose speed is αc.

In[55]:= $v = \dfrac{1}{137} \, c$; p = [**electron** PARTICLE] [*mass*] v;

$$N\left[\text{UnitConvert}\left[\dfrac{\hbar}{p}, \text{nm}\right], 1\right]$$

Out[55]= 0.3 nm

The de Broglie wavelength of an outer electron in an atom gives the size of the atom. Since outer electrons have a limited range energy of a few eV (and therefore, a limited momentum range), all atoms are roughly the same size.

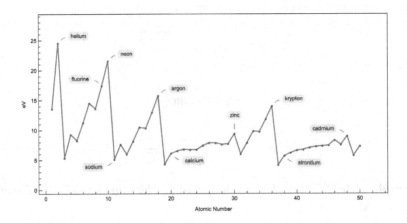

Figure 1.3 The atomic ionization energy is shown *vs.* atomic number.

Example 1.44 Calculate the de Broglie wavelength in fm of a neutron whose kinetic energy is 10 MeV.

In[56]:= m = [**neutron** PARTICLE] [*mass*] ; K = 10 MeV ; p = $\sqrt{2\,m\,K}$;

$$N\left[\text{UnitConvert}\left[\dfrac{h}{p}, \text{fm}\right], 1\right]$$

Out[57]=

9. fm

The de Broglie wavelength of a nucleon inside the nucleus gives the size of the nucleus.

Example 1.45 Calculate the de Broglie wavelength of a quark whose momentum is 200 MeV/c.

In[58]:= $\mathbf{p = 200 \dfrac{MeV}{c}}$; $\mathbf{N\left[UnitConvert\left[\dfrac{h}{p}, fm\right], 1\right]}$

Out[58]=

6. fm

The de Broglie wavelength of a quark inside a proton gives the size of the proton.

1.5.2 Wavelength and Kinetic Energy

The relationship between momentum and kinetic energy ,

$$K = E - mc^2 = \sqrt{(pc)^2 + (mc^2)^2} - mc^2,$$

gives

$$pc = \sqrt{K^2 + 2mc^2K},$$

and

$$\lambda = \frac{hc}{pc} = \frac{hc}{\sqrt{K + 2mc^2K}}$$

Example 1.46 Calculate the wavelength of an electron that has a kinetic energy of 2 MeV.

In[59]:= $\mathbf{K = 2\ MeV}$; $\mathbf{m = \boxed{electron\ \text{PARTICLE}}\boxed{mass}}$;

$\mathbf{ScientificForm\left[UnitConvert\left[\dfrac{h\ c}{\sqrt{K^2 + 2\ m\ c^2\ K}}\right], 3\right]}$

Out[60]//ScientificForm=

5.04×10^{-13} m

1.5.3 Classical Regime

In the regime of classical physics, de Broglie wavelengths are unmeasurably small.

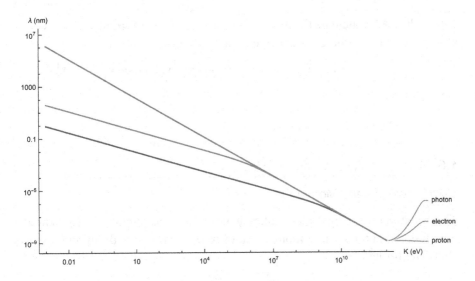

Figure 1.4 The de Broglie wavelengths of a photon, electron, and proton are shown vs. kinetic energy.

Example 1.47 Calculate the de Broglie wavelength of the earth in its orbit about the sun.

In[61]:= p = [**Earth** PLANET] [*mass*] [**Earth** PLANET] [*average orbit velocity*] ;

$$N\left[\text{UnitConvert}\left[\frac{h}{p}\right], 1\right]$$

Out[61]= $4. \times 10^{-63}$ m

Example 1.48 Calculate the de Broglie wavelength of a 5 gram snail moving at 0.03 miles per hour.

In[62]:= p = $\left(5\ g\right)\left(\theta.\theta3\ mi/h\right)$; NumberForm$\left[\text{UnitConvert}\left[\frac{h}{p}\right], 1\right]$

Out[62]//NumberForm=

$1. \times 10^{-29}$ m

Even if we assign a ridiculously small speed to the snail, its wavelength is tiny.

Example 1.49 Calculate the de Broglie wavelength of a 5 gram snail moving at a speed of one atomic distance per year.

In[63]:= $p = (5 \text{ g}) (0.3 \text{ nm} / \text{yr}); \text{NumberForm}\left[\text{UnitConvert}\left[\frac{h}{p}\right], 1\right]$

Out[63]//NumberForm=

$1. \times 10^{-14} \text{ m}$

1.6 UNCERTAINTY PRINCIPLE

1.6.1 Position and Momentum

The wave nature of particles makes it impossible to simultaneously know both the location (Δx) and momentum (Δp) of a particle with infinite precision. The limit is

$$\Delta x \Delta p \geq \frac{\hbar}{2}.$$

The conceptual reasoning is that the act of observing the position necessitates an interaction with at least one photon of wavelength equal to the position accuracy ($\lambda \sim \Delta x$), and the collision of that photon produces an uncertainty in momentum ($\Delta p \sim \frac{h}{\lambda}$), resulting in

$$\Delta x \Delta p \sim h.$$

Example 1.50 Calculate the minimum uncertainty in momentum for a particle whose position is known to 0.1 nm.

In[64]:= $\Delta x = .1 \text{ nm}; \text{NumberForm}\left[\Delta p = \text{UnitConvert}\left[\frac{\hbar}{2 \Delta x}\right], 1\right]$

Out[64]//NumberForm=

$5. \times 10^{-25} \text{ kg m/s}$

Note that the calculation of momentum above does not depend on the particle mass or whether or not the particle was relativistic. Suppose the confined particle is an electron.

Example 1.51 Calculate the minimum kinetic energy of an electron whose momentum is equal to Δp as calculated above, corresponding to $\Delta x = 0.1$ nm.

In[65]:= `NumberForm[UnitConvert[` $\dfrac{\Delta p^2}{2 \; \boxed{\text{electron } \text{PARTICLE}} \; \boxed{\textit{mass}}}$ `, ev], 1]`

Out[65]//NumberForm=

1. eV

The use of the nonrelativistic form of kinetic energy is seen to be justified since $K \ll mc^2$ (Ex. 1.22).

1.6.2 Energy and Time

A second form of the uncertainty principle says that

$$\Delta E \Delta t \geq \frac{\hbar}{2}.$$

Suppose (as once thought) that the strong force was mediated by exchange of a massive particle m whose range is governed by the uncertainty principle. The minimum energy of the particle is

$$\Delta E = mc^2$$

and the maximum range is

$$c\Delta t \approx 1 \text{ fm.}$$

Example 1.52 Use the uncertainty principle to estimate the mass energy of the exchanged particle.

In[66]:= `N[UnitConvert[` $\dfrac{\hbar \; c}{2 \left(1 \text{ fm}\right)}$ `, MeV], 1]`

Out[66]=

$1. \times 10^2$ MeV

In beta decay, weak force is mediated by the W particle whose large mass gives the force an extremely short range, even compared to that of the strong force.

Example 1.53 Get the W boson mass.

In[67]:= `Quantity["WBosonMass"]`

$\boxed{\text{W- boson } \text{PARTICLE}} \; \boxed{\textit{mass}}$

Out[67]=

m_W

Example 1.54 Use the uncertainty principle to estimate the range of the weak force.

In[69]:= m = [**W- boson** PARTICLE] [*mass*];

$$N\left[UnitConvert\left[\frac{\hbar \, c}{m \, c^2}\right], 1\right]$$

Out[70]=

$2. \times 10^{-18}$ m

CHAPTER 2

Thermal Radiation

2.1 RAYLEIGH-JEANS FORMULA

In its classical interpretation, radiation comes from atomic oscillators having a continuous probability distribution, which is distributed exponentially. The exponential distribution has a number of remarkale properties.

Example 2.1 Calculate the integral, average, and root-mean-square (rms) deviation from the average for the exponential distribution e^{-x}.

In[1]:= $\left\{ \int_0^\infty e^{-x} \, dx, \int_0^\infty x \, e^{-x} \, dx, \sqrt{\int_0^\infty (x-1)^2 \, e^{-x} \, dx} \right\}$

Out[1]= $\{1, 1, 1\}$

2.1.1 Average Oscillator Energy

To get the total radiation energy density, one needs to count the number of oscillators and then multiply by the average energy per oscillator. The average energy comes from an exponential probability distribution. For equilibrium at temoperature T, the probability per energy $f(E)$ is written

$$f(E) = Ce^{-\frac{E}{kT}} \, dE.$$

where k is the Boltzmann constant that converts temperature into energy, and C is a normalization constant making unit probability after summing over all energies.

Example 2.2 Get the Boltzmann constant.

In[2]:= `Quantity["BoltzmannConstant"]`

Out[2]= k

DOI: 10.1201/9781003395515-2

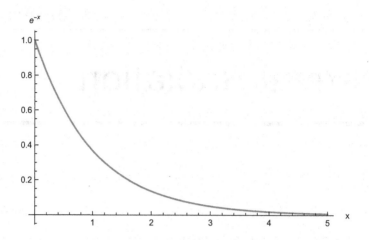

Figure 2.1 The exponential distribution, e^{-x}, has the remarkable property the the integral, average, and rms deviation from the average are all unity.

Example 2.3 Calculate the average energy per atomic oscillator.

In[3]:= $Assumptions = {k T > 0};$ $\dfrac{\int_0^\infty C\, E\, e^{-\frac{E}{kT}}\, dE}{\int_0^\infty C\, e^{-\frac{E}{kT}}\, dE}$

Out[3]= k T

In Ex. 2.3, the Greek letter E is used to avoid a reserved name (A.4), and k is entered as a user-defined variable (not the unit of Ex. 2.2) to avoid substitution of a number in the output.

An exponential distribution of oscillators gives an average energy of kT.

Example 2.4 Calculate the approximate value of kT at 290 K.

In[4]:= T = 290. K ; Rationalize$\left[\text{UnitConvert}\left[k\ T, ev\right], 10^{-3}\right]$

Out[4]= $\dfrac{1}{40}$ eV

It is handy to remember that at room temperature,

$$kT \approx \frac{1}{40}\text{eV}.$$

In Ex. 2.4, k is used as the unit.

2.1.2 Number of Modes

The next step is to count the number of oscillation modes. Consider a standing wave in one dimension of length L and wavelength λ. For 3 dimensions, consider a spherical volume $\frac{4}{3}\pi L^3$ and divide by the volume of one mode, λ^3. This gives

$$N = 2\frac{\frac{4}{3}\pi L^3}{\lambda^3},$$

where the factor of 2 accounts for the 2 polarizations of electromagnetic waves. The number of states per L^3, $n_s(\lambda)$, is

$$n_s = \frac{N}{L^3} = \frac{8\pi}{3\lambda^3}.$$

The number per volume per unit wavelength, referred to as the density of states $n(\lambda)$, is

$$n(\lambda) = -\frac{dn_s}{d\lambda} = \frac{8\pi}{\lambda^4}.$$

The energy per volume per wavelength stored in the cavity is the density of states times the average energy.

$$u(\lambda) = n(\lambda)kT = \frac{8\pi kT}{\lambda^4}.$$

2.1.3 Power per Area

There is one more thing to be done which is to uncover the relationship between energy per volume in a cavity and power per area radiated from its surface. Consider radiation at an angle θ to some surface (Fig. 2.2). The power radiated per wavelength (R) is related to the energy per volume per wavelength (u) by

$$R = \frac{1}{2}\frac{uV}{\Delta t} = \frac{1}{2}\frac{uLA\cos\theta}{L/(c\cos\theta)} = \frac{1}{2}cuA\cos^2\theta),$$

where the 1/2 comes from only one-half of the radiation in the cavity moving in a direction away from the wall. Averaging over all angles gives another factor of 1/2. Thus, the fundamental relationship between volume energy and power radiated from the surface is

$$R = \frac{c}{4}u.$$

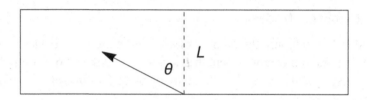

Figure 2.2 Radiation at an angle θ from a cavity wall. The time for radiation to cross the cavity is $L/(c\cos\theta)$ and the effective area from which the radiation originates is reduced by a factor $\cos\theta$ (all the radiation makes it across when $\theta = 0$ and none makes it across when $\theta = \frac{\pi}{2}$).

Example 2.5 Calculate the average value of $\cos^2\theta$ from $-\frac{\pi}{2}$ to $\frac{\pi}{2}$.

$$\text{In[5]:} \quad \frac{1}{\pi} \int_{-\frac{\pi}{2}}^{\frac{\pi}{2}} \text{Cos}[\theta]^2 \, d\theta$$

$$\text{Out[5]=} \quad \frac{1}{2}$$

The power per area per wavelength radiated from the surface is

$$R(\lambda) = \frac{c}{4}u(\lambda) = \frac{2\pi ckT}{\lambda^4}.$$

This is the famous Rayleigh-Jeans formula (Fig. 2.3). It only works in the limit of long wavelengths, suffering from what is referred to as the "ultraviolet catastrophe", as it blows up when $\lambda \to 0$. Solution of this problem was the birth of modern physics at a time when it was once widely thought that that all of fundamental physics was known!

The power per area due to photon wavelengths in an interval

$$\lambda_1 < \lambda < \lambda_2$$

may be obtained by direct integration (area under the curve of Fig. 2.3).

Example 2.6 Calculate the power per area radiated for $1\text{ m} < \lambda < 2\text{ m}$ at 300 K.

$$\text{In[6]:=} \quad T = 300 \text{ K}; \text{NumberForm}\left[\text{UnitConvert}\right.$$

$$\left[\int_{1.\text{ m}}^{2.\text{ m}} \frac{2\pi c k T}{\lambda^4} \, d\lambda, \frac{W}{m^2}\right], 3\right]$$

Out[6]//NumberForm=

$$2.28 \times 10^{-12} \text{ W/m}^2$$

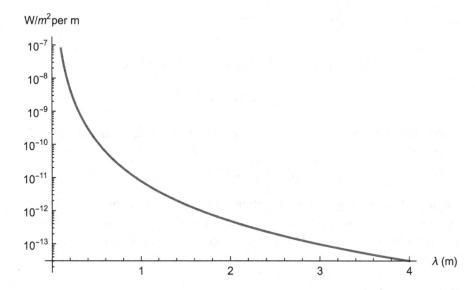

Figure 2.3 Radiated power per area per wavelength *vs.* λ predicted by the Rayleigh-Jeans formula at 300 K.

This result is reasonable but at smaller wavelengths something goes badly wrong.

Example 2.7 Calculate the power per area radiated for 400 nm < λ < 700 nm at 300 K.

In[7]:= **T = 300 K ;**

$$N\left[\text{UnitConvert}\left[\int_{400\ \text{nm}}^{700\ \text{nm}} \frac{2\pi c\ k\ T}{\lambda^4}\ d\lambda, \frac{W}{m^2}\right], 3\right]$$

Out[8]= $3.31 \times 10^7\ \text{W}/\text{m}^2$

This is clearly not correct: objects are not glowing from radiation in the visible spectrum at room temperature!

2.2 PLANCK FORMULA

The small-wavelength correction to the Rayleigh-Jeans formula was obtained by Planck who assumed that the atomic oscillators that produce the radiation do not have a continuous distribution of energy levels, but that they were quantized with energies

$$E_n = n\hbar\omega.$$

A photon emitted from the transition between adjacent levels has an energy

$$E_p = E_{n+1} - E_n = \hbar\omega = \frac{hc}{\lambda}.$$

The levels still have an exponential drop with energy, but are quantized with a spacing

$$\Delta E = E_{n+1} - E_n = \frac{hc}{\lambda}.$$

Figure 2.4 shows a quantized probability distribution with energy spacing of $\frac{kT}{2}$. The average energy has been reduced to $0.77\ kT$ compared to kT for a continuous distribution. The average energy drops rapidly with increased spacing.

Probability distribution

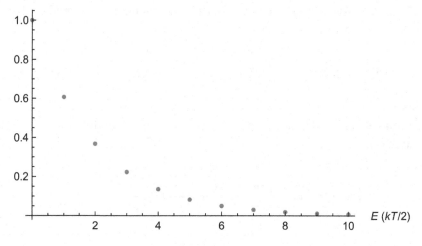

Figure 2.4 Exponential energy distribution for quantized energy levels with spacing $kT/2$.

Example 2.8 Calculate the average energy per oscillator for a quantized distribution with $\frac{kT}{2}$ energy spacing.

In[9]:= $N\left[\sum_{n=0}^{\infty} \frac{1}{2}\, n\, kT\, e^{-\frac{n}{2}} \bigg/ \sum_{n=0}^{\infty} e^{-\frac{n}{2}}\right]$

Out[9]= 0.770747 kT

Example 2.9 Calculate the average energy per oscillator for a quantized distribution with $10kT$ energy spacing.

In[10]:= $N\left[\sum_{n=0}^{\infty} 10\, n\, kT\, e^{-10\,n} \Big/ \sum_{n=0}^{\infty} e^{-10\,n}\right]$

Out[10]=
$\qquad$ 0.00045402 kT

At room temperature, this corresponds to about a factor of 2000 lower at $\lambda = 5 \times 10^{-6}$ m.

Example 2.10 Calculate the average energy per oscillator in terms of λ for energy spacing hc/λ.

In[11]:= **Clear[k, T]; Simplify$\left[\dfrac{\sum_{n=0}^{\infty} \dfrac{n\,(hc)\,e^{-\frac{nhc}{\lambda kT}}}{\lambda}}{\sum_{n=0}^{\infty} e^{-\frac{nhc}{\lambda kT}}}\right]$**

Out[11]= $\dfrac{c\,h}{\left(-1 + e^{\frac{ch}{kT\lambda}}\right)\lambda}$

Note that Mathematica chooses the order of the symbols in the output.

The famous universal thermal (blackbody) radiation formula derived by Planck is obtained by replacing kT in the Rayleigh-Jeans formula (2.1.3) with the average energy per oscillator from Ex. 2.10, giving

$$R = \frac{2\pi\hbar c^2}{\lambda^5(e^{\frac{hc}{\lambda kT}} - 1)}.$$

This is the correct answer for the observed power per area per wavelength radiated by an object in thermal equilibrium at a temperature T.

2.2.1 Comparison with Rayleigh-Jeans

At large λ the blackbody formula reduces to the Rayleigh-Jeans formula.

Example 2.11 Calculate the leading term in the blackbody formula at large λ.

In[12]:= **Series** $\left[\dfrac{2 \pi h c^2}{\lambda^5 \left(e^{\frac{c h}{k T \lambda}} - 1 \right)}, \{\lambda, \infty, 4\} \right]$

Out[12]= $\dfrac{T \left(2 \pi k c \right)}{\lambda^4} + O\left[\dfrac{1}{\lambda} \right]^5$

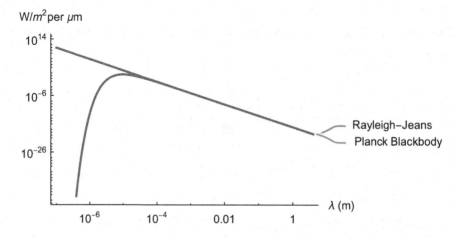

W/m^2 per μm

Figure 2.5 The Rayleigh-Jeans (upper) is compared to the Planck blackbody distribution (lower) at 300 K.

2.2.2 Planck Formula *vs.* Temperature

At higher temperatures, the total power radiated increases rapidly and peaks at lower wavelengths. The temerature 300 K (5800 K) is comparable to the surface of the earth (sun).

2.2.3 Radiation Peak

Plotting the blackbody radiation on a linear scale makes it easier to visualize the location of the peak.

The location of the peak of the distribution is found by setting the derivative equal to zero. The calculation is done over the domain of real numbers.

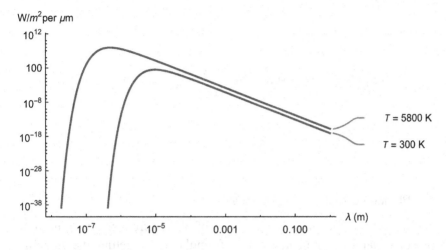

Figure 2.6 The blackbody distribution is shown *vs.* λ at 300 K and 5800 K.

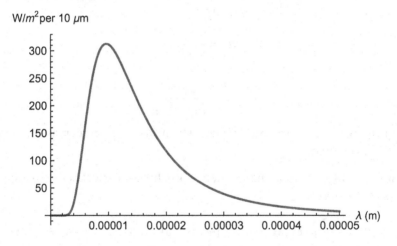

Figure 2.7 The blackbody distribution is shown *vs.* λ at 300 K on a linear scale.

Example 2.12 Get the symbol for the real domain.

In[13]:= **Reals**

Out[13]=

$\mathbb{R}$

Example 2.13 Calculate the wavelength at which the blackbody distribution peaks at $T = 300$ K.

In[14]:= $T = 300 \text{ K} ; L = \dfrac{h\ c}{k\ T} ;$

$\text{ScientificForm}\left[\text{NSolve}\left[D\left[\dfrac{1}{\lambda^5\ (e^{L/\lambda} - 1)}, \lambda\right] == 0, \lambda, \text{Reals}\right], 3\right]$

Out[15]//ScientificForm=

$\left\{\left\{\lambda \to 9.66 \times 10^{-6}\ m\right\}\right\}$

2.2.4 Planck Formula *vs.* Frequency

To get the radiation formula as a function of photon frequency (f), make a change of variables to the entire Planck formula, not forgetting the differential,

$$\lambda \to \frac{c}{f},$$

$$d\lambda = -\frac{c}{f^2}df,$$

and

$$R(f) = \frac{2\pi h f^3}{c^2(e^{\frac{hf}{kT}} - 1)}.$$

The negative sign in the differential, which just says f increases as λ decreases, is ignored.

Example 2.14 Calculate the frequency at which the blackbody distribution peaks at $T = 300$ K.

In[16]:= $f_\theta = \dfrac{k\ T}{h} ; \text{ScientificForm}\left[\text{NSolve}\left[D\left[\dfrac{f^3}{\left(e^{\frac{f}{f_\theta}} - 1\right)}, f\right] == 0, f, \text{Reals}\right], 3\right]$

Out[16]//ScientificForm=

$\left\{\left\{f \to 1.76 \times 10^{13}\ Hz\right\}\right\}$

2.2.5 Number of Photons

To get the photon flux (number per s per m^2), convert the freqeuncy scale to energy ($E = hf$) and divide by the photon energy.

$$F(E) = \frac{1}{E}R(E) = \frac{2\pi E^2}{(hc)^2(e^{\frac{E}{kT}} - 1)}$$

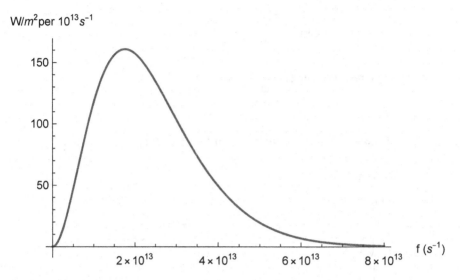

Figure 2.8 The blackbody distribution is shown *vs.* f at 300 K on a linear scale.

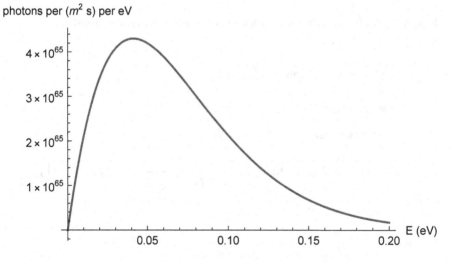

Figure 2.9 The photon flux is shown *vs.* energy at 300 K.

Example 2.15 Calculate the flux of visible photons at $T = 300$ K. A variable kT is defined to facilitate the numerical integration.

In[17]:= $E_1 = \text{UnitConvert}\left[\dfrac{h\,c}{700\ \text{nm}}\ \dfrac{1}{\text{eV}}\right];$

$E_2 = \text{UnitConvert}\left[\dfrac{h\,c}{400\ \text{nm}}\ \dfrac{1}{\text{eV}}\right];$

$T = 300\ \text{K};\ kT = \dfrac{1}{\text{eV}}\ \text{UnitConvert}\left[k\,T,\ \text{eV}\right];$

$\text{NumberForm}\Big[$

$\quad\text{UnitConvert}\Big[$

$\qquad \left(\dfrac{2\,\pi}{h^2\,c^2}\right)\ \dfrac{\text{J eV}}{\text{s}}\ \text{NIntegrate}\left[\dfrac{E^2}{\left(e^{E/kT} - 1\right)},\ \{E,\ E_1,\ E_2\}\right]\Big],$

$\quad 2\Big]$

$\text{Out[19]//NumberForm=}$

$\qquad 3.7\,\text{per meter}^2\ \text{per second}$

Example 2.16 Calculate the flux of visible photons at $T = 2000$ K.

In[20]:= $T = 2000\ \text{K};$

$kT = \dfrac{1}{\text{eV}}\ \text{UnitConvert}\left[k\,T\right];$

$\text{NumberForm}\Big[$

$\quad\text{UnitConvert}\Big[$

$\qquad \left(\dfrac{2\,\pi}{h^2\,c^2}\right)\ \dfrac{\text{J eV}}{\text{s}}\ \text{NIntegrate}\left[\dfrac{E^2}{\left(e^{E/kT} - 1\right)},\ \{E,\ E_1,\ E_2\}\right]\Big],$

$\quad 2\Big]$

$\text{Out[20]//NumberForm=}$

$\qquad 5.8 \times 10^{26}\ \text{per meter}^2\ \text{per second}$

2.3   STEFAN-BOLTZMANN LAW

If we integrate the power per area per wavelength over all wavelengths, the answer depends on only one parameter, the temperature, which appears to the fourth power.

Example 2.17 Integrate the thermal radiation formula over all wavelengths to get the total radiated power per area.

$Assumptions = L > 0; R = 2 π h c ^{2} $\int_0^\infty \dfrac{1}{\lambda^5 (e^{L/\lambda} - 1)}$ dλ /. L $\to$ $\dfrac{h\ c}{k\ T}$

Out[22]=

$$T^4 \left(\dfrac{2\,\pi^5}{15}\ k^4 / (h^3 c^2) \right)$$

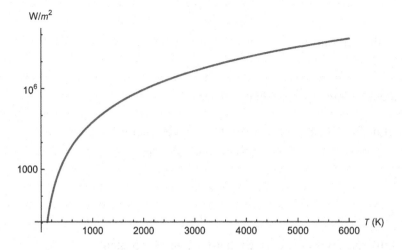

Figure 2.10 The total radiated power per area is shown *vs.* temperature.

Example 2.18 Find the temperature where the radiated power per area is 1 MW/m^2.

In[23]:= $Assumptions = T > 0; N$\left[\text{Solve} \left[T^4\ \dfrac{2\,\pi^5\ k^4}{15\ h^3\ c^2} == \dfrac{10^6\ W}{m^2}, T \right], 3 \right]$.{1}

Out[23]=

$$\left\{ T \to 2.05 \times 10^3\ K \right\}$$

Example 2.19 Find the value of kT where the radiated power per area is 1 MW/m^2.

In[24]:= **$Assumptions = kT > 0;**

$$\text{sol = N}\left[\text{Solve}\left[\text{kT}^4 \; \frac{2\,\pi^5}{15\; h^3\; c^2} \; == \; \frac{10^6\; W}{m^2}\,, \; \text{kT}\right], \; 3\right].\{1\};$$

UnitConvert[kT /. sol, eV]

Out[25]= 0.177 eV

The Stefan-Boltzmann law for the radiated power per area is written

$$\frac{P}{A} = \sigma T^4$$

where the constant

$$\sigma = \frac{2\pi^5 k^4}{15h^3 c^2}$$

is called the Stefan-Boltzmann constant.

Example 2.20 Get the Stefan-Boltzmann constant.

In[26]:= **Quantity["StefanBoltzmannConstant"]**

Out[26]=

σ

Example 2.21 Calculate the Stefan-Boltzmann constant.

In[27]:= **σ = ScientificForm$\left[\text{UnitConvert}\left[\frac{2.\;\pi^5\; k^{\,4}}{15\; h^{\,3}\; c^{\,2}}\,, \; \frac{W}{m^{\,2}\; K^{\,4}}\right], \; 3\right]$**

Out[27]//ScientificForm=

$5.67 \times 10^{-8}\; W/\, (m^2 K^4)$

It is may be useful to express this in eV, for example, if we are counting photons.

Example 2.22 Calculate the Stefan-Boltzmann constant using eV for energy units.

In[28]:= **N$\left[\text{UnitConvert}\left[\frac{2\,\pi^5\; k^{\,4}}{15\; h^{\,3}\; c^{\,2}}\,, \; \frac{eV}{s\; m^{\,2}\; K^{\,4}}\right], \; 3\right]$**

Out[28]=

$3.54 \times 10^{11}\; eV/\, (m^2 s\; K^4)$

It may be useful to factor out k^4 so we have something (σ') times $(kT)^4$.

$$\frac{P}{A} = \sigma'(kT)^4$$

Example 2.23 Calculate σ'.

In[29]:= $\sigma' = N\left[\text{UnitConvert}\left[\dfrac{2\,\pi^5}{15\ h^3\ c^2}\,,\ \dfrac{W}{m^2\ eV^4}\right],\ 3\right]$

Out[29]=

$1.03 \times 10^9\ W/\,(m^2 eV^4)$

Example 2.24 Calculate the radiated power per area for $kT = \frac{1}{40}$ eV.

In[30]:= $\sigma'\left(\dfrac{1}{40}\ eV\right)^4$

Out[30]=

$402.\ W/m^2$

Example 2.25 Find the total radiated power per area from an object at room temperature.

In[31]:= $T = 300\ K\ ;$

$N\left[\text{UnitConvert}\left[\sigma\ T^4,\ \dfrac{W}{m^2}\right],\ 3\right]$

Out[32]=

$459.\ W/m^2$

Example 2.26 Get the surface temperature of the sun.

In[33]:= $T = \text{UnitConvert}\left[\boxed{\textbf{Sun}\ \text{STAR}}\boxed{\textit{effective temperature}},\ K\right]$

Out[33]=

$5772.\ K$

Example 2.27 Calculate the power per area radiated from sun.

In[34]:= $N\left[\text{UnitConvert}\left[R,\ \dfrac{W}{m^2}\right],\ 3\right]$

Out[34]=

$6.29 \times 10^7\ W/m^2$

Example 2.28 Calculate the total power radiated from sun.

In[35]:= $\text{P} = \text{N}\Big[\text{UnitConvert}\Big[\text{R}\,4\,\pi\,\boxed{\textbf{Sun}\ \text{STAR}}\,\Big[\boxed{\textit{average radius}}\Big]^{2},\ \text{W}\Big],\ 3\Big]$

Out[35]=

$3.83 \times 10^{26}\,\text{W}$

2.4 WEIN'S LAW

The peak wavelength value of the thermal spectrum is found by taking its derivative with respect to λ and setting it equal to 0. This results in a transcendental equation which can be solved numerically with the use of the function NSolve [ ]. The peak value occurs at about $\frac{hc}{5kT}$.

Example 2.29 Find λ that gives the peak radiated power per area.

In[36]:= `ClearAll["Global`*"];`

$\text{sol} = \text{NSolve}\Big[\text{D}\Big[\dfrac{1}{\lambda^{5}\,\big(e^{L/\lambda}-1\big)},\ \lambda\Big] == 0,\ \lambda,\ \text{Reals}\Big]\ /.\ \text{L} \to \dfrac{h\,c}{k\,T};$

$\lambda_{max} = \lambda\ /.\ \text{sol}[\![1]\!]$

Out[37]=

$\dfrac{0.201405\,h\,c\,/\,k}{T}$

This is expression commonly written as λT equals a constant.

Example 2.30 Find the value of λT which gives the wavelength for maximum power.

In[38]:= `NumberForm[UnitConvert[`λ_{max}` T], 4]`

Out[38]//NumberForm=

$0.002898\,\text{m}\,\text{K}$

This is Wein's law. The peak wavelength is given by

$$\lambda_{max} T = 0.002898\ \text{m} \cdot \text{K}.$$

Example 2.31 Find the wavelength for maximum power from the sun.

In[39]:= **NumberForm**$\left[\text{UnitConvert}\left[\dfrac{\lambda_{max}\ T}{\boxed{\textbf{Sun}\ \text{STAR}}\ \left[\ \boxed{effective\ temperature}\ \right]}\ ,\ nm\ \right],\ 3\right]$

Out[39]//NumberForm=

    502. nm

The human eye has evolved to have sensitivity at the wavelength where the sun puts out its peak power.

CHAPTER 3

Key Processes

3.1 RADIOACTIVE DECAY

Radioactive decay produced the mystery particles α, β, and γ (He nucleus, electron, photon) and played a central role at the beginning of the modern physics era. Study of these decays revealed the structure of the atom at a time when α, β, and γ were still unknown radiation phenomena.

3.1.1 Decay Types

Decay always occur if they are not forbidden. Alpha decay is a common decay of a heavy nucleus. The process is

$$\frac{A}{Z}X \rightarrow \frac{A-2}{Z-2}Y + \alpha,$$

where the decaying nucleus has shed both 2 neutrons and 2 protons that are bound into the α particle.

Beta decay has two types, where either a neutron turns into a proton inside the nucleus,

$$\frac{A}{Z}X \rightarrow \frac{A}{Z+1}Y + \beta^+ + \nu_e,$$

or a proton turns into a neutron inside the nucleus,

$$\frac{A}{Z}X \rightarrow \frac{A}{Z-1}Y + \beta^- + \bar{\nu}_e,$$

where the β^- (β^+) particle is an electron (positron).

Related to β decay is the process of electron capture,

$$\frac{A}{Z}X + \beta^- \rightarrow \frac{A}{Z-1}Y + \nu_e,$$

where a nucleus has captured an inner atomic electron and turned a proton into a neutron.

DOI: 10.1201/9781003395515-3

Gamma decay occurs when a nucleus has been left in an excited state (X^*),

$$\,_Z^A X^* \rightarrow \,_Z^A X + \gamma.$$

3.1.2 Decay Probability

Radioactive decays are governed by an exponential distribution (see 2.1),

$$N(t) = N_0 e^{-t/\tau},$$

where $N(t)$ represents the size of the sample as a function of time, N_0 is the sample size at time $t = 0$, and τ is the average lifetime.

Example 3.1 Calculate the average lifetime.

In[1]:= $Assumptions = τ > 0$; $\left(\int_0^\infty t\, e^{-t/\tau}\, dt \right) \Big/ \int_0^\infty e^{-t/\tau}\, dt$

Out[1]= τ

The exponential distribution has the special property that the rms deviation from the mean is equal to the mean.

Example 3.2 Calculate the rms deviation from the mean lifetime.

In[2]:= Simplify$\left[\sqrt{ \left(\int_0^\infty (t - \tau)^2\, e^{-t/\tau}\, dt \right) \Big/ \int_0^\infty e^{-t/\tau}\, dt } \,\right]$

Out[2]= τ

The half-life $(t_{1/2})$ is defined by

$$\frac{1}{2} = e^{-t_{1/2}/\tau}$$

which gives

$$t_{1/2} = \tau \ln 2.$$

Uranium-238 decays by α emission,

$$^{238}\text{U} \rightarrow \,^{234}\text{Th} + \alpha.$$

Example 3.3 Calculate the initial decay rate in a 1 g sample of U-238.

In[3]:= $\tau = \dfrac{\boxed{\text{uranium-238 ISOTOPE}}\ \boxed{\big[\,\boxed{\textit{half-life}}\,\big]}}{\text{Log[2]}}$;

$R = \dfrac{N_{\theta}}{238}\left(1 - e^{-1\ s/\tau}\right)\ s^{-1};$

ScientificForm[UnitConvert[R], 3]

Out[4]//ScientificForm=

1.25×10^{4} per second

Thorium-234 decays by β emission,

$$^{234}\text{Th} \rightarrow{}^{234}\text{Pa} + e^{-} + \bar{v}_{e}.$$

Example 3.4 Calculate the initial decay rate of Th-234.

In[5]:= $\tau = \dfrac{\boxed{\text{thorium-234 ISOTOPE}}\ \boxed{\big[\,\boxed{\textit{half-life}}\,\big]}}{\text{Log[2]}}$;

UnitConvert$\left[\dfrac{R\,(1\ s)}{\tau}\right]$

Out[6]= 0.00415 per second

In[7]:= τ

Out[7]= 34.8 days

3.1.3 Carbon Dating

The isotope carbon-14 and normal carbon-12 are in atmospheric equilibrium in the ratio 1.2×10^{-12}. Living organisms contain this natural mix of ^{12}C and ^{14}C, but upon death no longer take in any more ^{14}C which then decays away with a half-life of 5730 years.

Example 3.5 An old bone has a ^{14}C fraction that has dropped by 20% from atmospheric equilibrium to 0.96×10^{-12}. How old is the bone?

```
In[8]:=  ClearAll["Global`*"]; $Assumptions = x ∈ ℝ;
         R = 1.2 × 10⁻¹²; r = 0.96 × 10⁻¹²;

                             5730 yr
         x = Solve[r == R e⁻ᵗ/τ, t] /. τ → ─────── ;
                              Log[2]

         NumberForm[x〚All, 1, 2〛.{1}, 3]
```

Out[11]//NumberForm=

    1840. yr

3.2 MOTION OF A CHARGED PARTICLE IN ELECTRIC AND MAGNETIC FIELDS

3.2.1 Charged Particle in an Electric Field

Newton's 2nd law for a charged particle in an electric field (E) is

$$F = qE = \frac{q\Delta V}{d},$$

where the field is given by a potential difference (ΔV) across a distance (d). This force is most conveniently expressed in eV/m (1.1.2).

Example 3.6 Calculate the force on an electron in an electric field of 100 V/cm in both eV/m and N.

```
                100 V
In[12]:=  F = e ─────── ;
                 cm


                       keV
          UnitConvert[F, ─── ]
                        m

          ScientificForm[N[UnitConvert[F, N ], 3]]
```

Out[13]=

       10 keV/m

Out[14]//ScientificForm=

    1.60 × 10⁻¹⁵ N

3.2.2 Charged Particle in a Magnetic Field

Newton's 2nd law for the circular motion of radius r for a particle with mass m and charge q in a magnetic field B is

$$F = m\frac{v^2}{r} = qvB,$$

which gives

$$mv = p = qrB.$$

If the electron is relativistic, the same equation, $p = qrB$, holds provided that the relativistic expression for momentum (1.3.3) is used.

Example 3.7 Find the radius of curvature of a 10 keV/c electron in a 1 mT magnetic field.

p = 10 keV/c; B = 0.001 T;

$$\text{ScientificForm}\left[\text{UnitConvert}\left[\frac{p}{e\,B}\right], 3\right]\|$$

Out[15]//ScientificForm=

$$3.34 \times 10^{-2} \text{ m}$$

3.2.3 Electron Charge-to-Mass Ratio

The charge-to-mass ratio of the electron was measured by J. J. Thompson who built the first mass spectrometer consisting of parallel plates having perpendicular electric and magnetic fields. The idea of the spectrometer is that an electron velocity is very difficult to measure but its path is relatively easy. With no magnetic field, the electron gets an upward acceleration (a) of

$$a = \frac{F}{m} = \frac{qE}{m}.$$

The upward component of the electron velocity (v_y) is

$$v_y = at = \frac{aL}{v_x},$$

where (L) is the distance traveled between the plates, t is the travel time, and v_x is the initial electron velocity. Thus,

$$\frac{q}{m} = \frac{v_x v_y}{EL}.$$

When the magnetic field is also applied with a strength that balances the electric force,

$$qE = qv_xB.$$

Thus, using

$$v_x = \frac{E}{B}$$

and

$$\tan\theta = \frac{v_y}{v_x},$$

the result may be written

$$\frac{q}{m} = \frac{E\tan\theta}{LB^2}.$$

In the above expression, E is the electric field that causes a deflection θ and B is the magnetic field that causes no deflection.

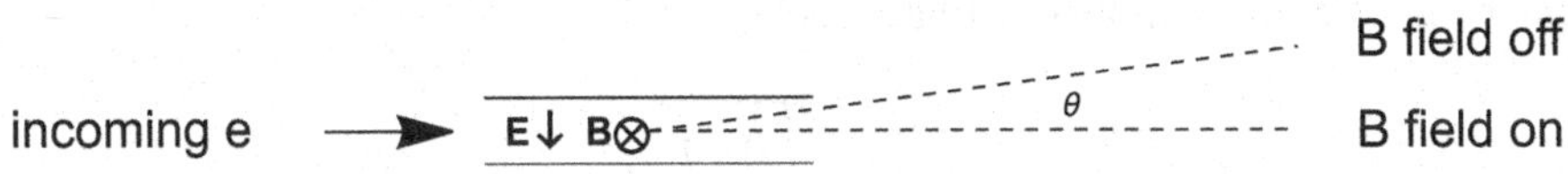

Figure 3.1 A region of space has an electric field in the downward direction and a magnetic field into the page. When the electric and magnetic forces cancel, an incoming electron is undeflected. When the magnetic field is turned off, the electron is deflected upward.

Example 3.8 Calculate the electron speed for a $\theta = 0.1$ (radian) deflection from an electric field of 10^4 V/m and $L = 5$ cm.

```
In[16]:=  E = 10^4  V/m ;

          L = 5 cm ;

          θ = .1;

          m = | electron  PARTICLE | [ | mass | ] ;

              ┌─────────
          B = │  E Tan[θ]
              │  ─────────   ;
              │   L  e/m
             √

          v = NumberForm[ UnitConvert[ E/B ], 3 ]
```

Out[16]//NumberForm=

2.96×10^7 m/s

This is an atomic speed (1.2.3); the electrons used for this measurement came from atoms.

3.2.4   Electron Charge Measurement

The electron charge was measured by R. Millikan in his famous oil-drop experiment. A charged droplet of mass m suspended in air reaches a terminal speed v_T given by the force balance

$$mg = bv_T = 6\pi\eta Rv_T,$$

where the constant b is proportional to both the viscosity of air (η) and the droplet radius (R) through a relationship called Stokes's Law. Using the formula for density (ρ),

$$\rho = \frac{m}{\frac{4}{3}\pi R^3},$$

the droplet radius can be calculated as

$$R = \sqrt{\frac{9v_T\eta}{2g\rho}}.$$

If an electric field is now applied in a direction to make the droplet move upward, the force balance is

$$qE = mg + bv_E,$$

where v_E is the electron velocity with the field on. This can be used to eliminate b to get

$$q = \frac{mg(1 + \frac{v_E}{v_T})}{E}.$$

Example 3.9 For $v_T = 1.3$ mm/s and $\rho = 0.9$g/cm^3, determine b, R, and m.

In[17]:= η = ThermodynamicData["Air", "Viscosity",

{"Temperature" → 20 °C, "Pressure" → 1 atm}];

$v_T = 1.3 \dfrac{mm}{s}$;

$\rho = 0.9 \dfrac{g}{cm^3}$; $R = \sqrt{\dfrac{9 \, v_T \, \eta}{2 \, g \, \rho}}$; $m = \dfrac{4}{3} \pi R^3 \rho$;

NumberForm[UnitConvert[R], 3]
NumberForm[UnitConvert[m], 3]

NumberForm$\left[\text{UnitConvert}\left[\dfrac{m\,g}{v_T}\right], 3\right]$

Out[19]//NumberForm=

3.47×10^{-6} m

Out[20]//NumberForm=

1.58×10^{-13} kg

Out[21]//NumberForm=

1.19×10^{-9} kg/s

The electron charge measurement combined with the charge-to-mass ratio measurement (3.2.3) gives the electron mass.

3.3 PHOTOELECTRIC EFFECT

The photoelectric effect (PE) occurs when a photon has sufficient energy to eject an electron from the metal. The electron which was originally bound has absorbed the photon and become free. The electron binding energy is referred to as the work function. The PE cannot occur if the photon energy is less that the binding energy. If the photon energy is greater than the binding energy, the freed electron gets the excess kinetic energy. The process is

$$\gamma + e \text{ (in metal)} \rightarrow e \text{ (free)}$$

Note that a free electron cannot absorb a photon. The metal is needed in this case to conserve momentum, but it is so heavy that it gets no energy.

Traditionally, the electron kinetic energy is measured by the voltage (V_0) needed to stop all emitted electrons, the so-called stopping voltage. The minimum binding energy is called the work function (ϕ) and an electron with

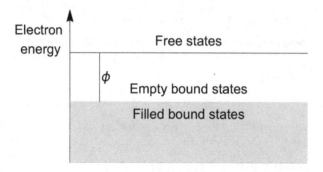

Figure 3.2 The energy diagram for the photoelectric effect shows empty and filled states.

binding energy ϕ gets the maximum allowed kinetic energy. Conservation of energy reads

$$eV_0 = K_{max} = hf - \phi$$

Example 3.10 Get the work function for lead.

In[22]:= `UnitConvert[` `lead` ELEMENT `[` `work function` `], eV ]`

Out[22]=

4.25 eV

Example 3.11 Get the work functions for elements 28 to 32.

In[23]:= `ionizationData =`
`    DeleteMissing[`
`      EntityValue[Take[EntityList["Element"], {28, 32}],`
`      {EntityProperty["Element", "AtomicNumber"],`
`        EntityProperty["Element", "WorkFunction"]},`
`      "EntityAssociation"]]`

Out[23]= ⟨| `nickel` → {28, (5.02 to 5.37) eV},

`copper` → {29, (4.46 to 5.12) eV},

`zinc` → {30, (3.62 to 4.92) eV},

`gallium` → {31, 4.32 eV}, `germanium` → {32, 5.00 eV} |⟩

The threshold frequency (f_0) below which there are no electrons emitted is given by

$$f_0 = \frac{\phi}{h}$$

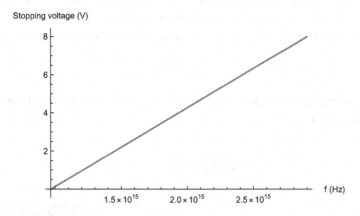

Figure 3.3 The stopping voltage is shown *vs.* radiation frequency.

Example 3.12 Calculate the threshold wavelength for the photoelectric effect in potassium.

In[24]:= ϕ = [**potassium** ELEMENT] [*work function*];

$$N\left[UnitConvert\left[\frac{h\ c}{\phi},\ nm\right],\ 3\right]$$

Out[25]=

541. nm

The PE occurs instantaneously. Classically, it could not occur until enough energy was absorbed.

Example 3.13 Photons with intensity 10^{-8} W/m^2 are incident on potassium. Estimate the amount of time classically that it would take an atom to absorb enough energy to eject an electron.

In[26]:= A = (.2 nm)^2; NumberForm$\left[UnitConvert\left[\dfrac{\phi}{A\left(10^{-8}\ \frac{W}{m^2}\right)}\right],\ 1\right]$

Out[26]//NumberForm=

9. $\times 10^8$ s

3.4 ELECTRON DIFFRACTION

3.4.1 Scattering Off a Crystal

When electrons are incident on a regular array of atoms, there are interference effects and the wave properties of the electrons are observable. In Fig. 3.4, θ is the angle between the incident electron direction and the planes of atoms indicated by the dashed lines. The condition for constructive interference is

$$n\lambda = 2d\sin\theta,$$

where d is the distance between layers of atoms (distance between the dashed lines in Fig. 3.4).

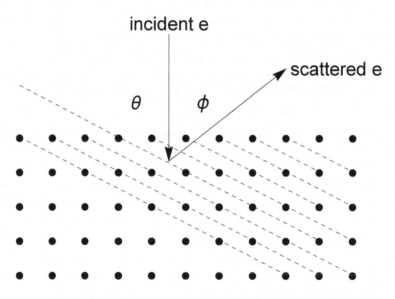

Figure 3.4 Electron scattering geometry from a crystal array of atoms.

The incident electrons are accelerated by a voltage V and have kinetic energy

$$K = eV = \frac{1}{2}mv^2.$$

In the original experiment of Davisson-Germer, peaks were observed at equally spaced intervals when plotted as the square-root of the acceleration voltage,

$$\sqrt{V} = v\sqrt{\frac{m}{2e}}.$$

The interference condition gives

$$n = \frac{2d \sin \theta}{\lambda},$$

thus, the data were suggesting that

$$\frac{1}{\lambda} \sim v.$$

In fact, precise agreement is given by the de Broglie formula (1.5),

$$\lambda = \frac{h}{p} = \frac{h}{mv} = \frac{h}{\sqrt{2meV}}.$$

Example 3.14 Calculate the wavelength of an electron accelerated through 100 V.

m = [**electron** PARTICLE] [*mass*];

V = 100 V; N[UnitConvert[$\dfrac{h}{\sqrt{2\,m\,e\,V}}$, nm], 2]

Out[27]= 0.12 nm

Direct observation of electron wave properties proved that the concept of de Broglie wavelength applies to the particle world.

3.4.2 Neutrons in Thermal Equilibrium

Nonrelativistic particles in thermal equilibrium have average kinetic energy

$$K = \frac{p^2}{2m} = \frac{3}{2}kT.$$

This gives

$$p = \sqrt{3mkT},$$

and

$$\lambda = \frac{h}{\sqrt{3mkT}}.$$

Example 3.15 Calculate the wavelength of thermal neutrons at 300 K and 77 K (liquid nitrogen temperature).

In[29]:= **m =** ⎡ **neutron** PARTICLE ⎤ ⎡ **mass** ⎤ ⎤ ;

$$N\left[\text{UnitConvert}\left[\frac{h}{\sqrt{3\,m\,k\,T}}, \text{nm}\right] /. T \to 300\,K, 2\right]$$

$$N\left[\text{UnitConvert}\left[\frac{h}{\sqrt{3\,m\,k\,T}}, \text{nm}\right] /. T \to 77\,K, 2\right]$$

Out[29]= 0.15 nm

Out[30]= 0.29 nm

3.4.3 Quarks Inside a Proton

The concept of de Broglie wavelength applies to quarks inside the protron, even though they are relativistic.

Example 3.16 Estimate the speed of a charm quark inside the J/ψ particle.

In[30]:= **m =** ⎡ **charm quark** PARTICLE ⎤ ⎡ **mass** ⎤ ⎤ ;

$$\lambda = 2 \text{ fm} ;$$

$$p = \frac{h}{\lambda} ;$$

$$E = \sqrt{\left(m\,c^2\right)^2 + \left(p\,c\right)^2} ;$$

$$v = N\left[\text{UnitConvert}\left[\frac{p\,c}{E}\right], 1\right] c$$

Out[31]=

0.4 c

3.5 COMPTON SCATTERING

Compton scattering originally referred to the scattering of a photon from an electron at rest. More generally it can refer to the scattering of a photon with any charged particle, whether moving or not. The electron gains kinetic energy in the collision and the photon loses kinetic energy.

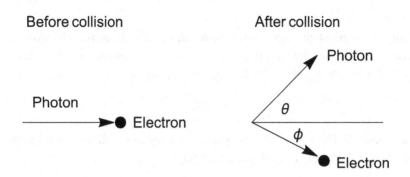

Figure 3.5 When a photon scatters off an electron at rest, the photojn and electron emerge at different angles.

3.5.1 Compton Formula in Terms of Energy

Let E_1 (E_2) be the incoming (outgoing) photon energy and p be the outgoing electron momentum. Conservation of energy gives

$$E_1 + mc^2 = E_2 + \sqrt{(pc)^2 + (mc^2)^2}$$

Conservation of momentum gives

$$\frac{E_1}{c} = p\cos\phi + \frac{E_2}{c}\cos\theta$$

$$p\sin\phi = \frac{E_2}{c}\sin\theta$$

Solve by squaring the square root in the energy equation and substituting for pc using the momentum equations.

Example 3.17 Solve for the photon energy.

```
In[33]:= ClearAll["Global`*"];
        Solve[ (E₁ + m c² - E₂)² - (E₁ - E₂ Cos[θ])² - (E₂ Sin[θ])² - (m c²)² == 0,
        E₂]

Out[33]= {{E₂ →  c² m E₁
                ─────────────────── }}
                c² m + E₁ - Cos[θ] E₁
```

3.5.2 Limiting Cases

When the incident photon energy is much smaller than the electron mass energy, it is like scattering off a "brick wall". Momentum is transferred but essentially no energy is lost by the incoming photon.

$$E_1 = E_2$$

Example 3.18 Find the scattered photon energy at small values (compared to the electron rest energy) of incident energy.

In[34]:= $\text{Series}\left[\dfrac{c^2\, m\, E_1}{c^2\, m + E_1 - \text{Cos}[\theta]\, E_1},\ \{E_1,\ 0,\ 1\}\right]$

Out[34]= $E_1 + O[E_1]^2$

For forward scattering ($\theta = 0$), the photon energy is unchanged.

Example 3.19 Evaluate the Compton formula at $\theta = 0$.

In[35]:= $\dfrac{c^2\, m\, E_1}{c^2\, m + E_1 - \text{Cos}[\theta]\, E_1}\ /.\ \theta \to 0$

Out[35]= E_1

For backward scattering ($\theta = \pi$) in the limit of large incoming photon energy (greater than the electron mass energy), the outgoing photon gets $1/2$ of the electron mass energy.

Example 3.20 Evaluate the Compton formula at $\theta = \pi$ and large incoming photon energy.

In[36]:= $\text{Limit}\left[\dfrac{c^2\, m\, E_1}{c^2\, m + E_1 - \text{Cos}[\theta]\, E_1}\ /.\ \theta \to \pi,\ E_1 \to \infty\right]$

Out[36]= $\dfrac{c^2\, m}{2}$

3.5.3 Compton Formula in Terms of Wavelength

The standard way to describe the collision is to calculate the change in wavelength of the photon, the so-called Compton formula by using the formula for photon wavelength ($\frac{hc}{\lambda}$).

$$\Delta\lambda = \lambda_1 - \lambda_2 = \frac{hc}{mc^2}(1 - \cos\theta)$$

Example 3.21 Solve for the photon wavelength difference.

In[37]:= **Solve**$\left[\frac{hc}{\lambda_2} == \frac{c^2 m \frac{hc}{\lambda_1}}{c^2 m + \frac{hc}{\lambda_1} - \text{Cos}[\theta] \frac{hc}{\lambda_1}} \text{ \&\& } \Delta\lambda == \lambda_2 - \lambda_1, \{\Delta\lambda, \lambda_2\} \right].$

$\{1, 0\}$ // **Simplify**

Out[37]= $\left\{ \Delta\lambda \to \frac{hc - hc \, \text{Cos}[\theta]}{c^2 m} \right\}$

The quantity,

$$\lambda_C = \frac{hc}{mc^2},$$

is known as the Compton wavelength.

Example 3.22 Calculate the numerical value of the Compton wavelength.

In[38]:= $\lambda_c = $ **ScientificForm**$\left[\text{UnitConvert} \left[\frac{h \, c}{\boxed{\textbf{electron} \; \text{PARTICLE}} \boxed{\boxed{mass}} \; c^2} \right], \right.$

$\left. 3 \right]$

Out[38]//ScientificForm=
2.43×10^{-12} m

3.5.4 Relationship of λ_c to Other Quantities

The Bohr radius and the Rydberg constant are discussed in **??** and 5.5. The Compton wavelength compared to the Bohr radius (a) is

$$\lambda_c = 2\pi\alpha a$$

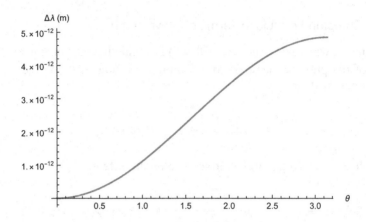

Figure 3.6 The change in photon wavelength depends on the scattering angle.

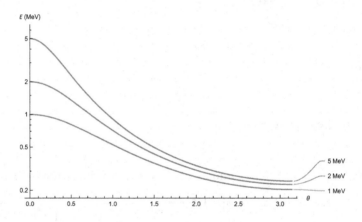

Figure 3.7 The energy of the scattered photon is shown *vs.* scattering angle for 1, 2, and 5 MeV incident photon energies.

Example 3.23 Calculate the Compton wavelength from the Bohr radius.

In[39]:= m = electron PARTICLE [mass];

$$\alpha = \frac{e^2}{4 \pi \, \varepsilon_\theta \, \hbar \, c};$$

$$a = \frac{4 \pi \hbar^2 \, \varepsilon_\theta}{e^2 \, m};$$

ScientificForm[UnitConvert[2 π α a], 3]

Out[40]//ScientificForm=

2.43×10^{-12} m

Example 3.24 Calculate the Compton wavelength from the Rydberg constant.

In[41]:= $R_\infty = \dfrac{\frac{1}{2} m (\alpha c)^2}{h\ c}$; ScientificForm$\left[\text{UnitConvert}\left[\dfrac{1}{2}\ \alpha^2\ \dfrac{1}{R_\infty}\right],\ 3\right]$

Out[41]//ScientificForm=

$\quad 2.43 \times 10^{-12}$ m

The classical electron radius (r_e)) is given by setting

$$\frac{e^2}{4\pi\varepsilon_0 r_e} = mc^2$$

which is the radius where the potential energy of an electron charge distribution would equal to its mass energy. It turns out that there is nothing special about this distance. The electron has no measurable structure (to date) and experiments have been performed at distances 5 orders of magnitude smaller than r_e. The Compton wavelength compared to the classical electron radius is

$$\lambda_c = 2\pi \frac{r_e}{a}$$

Example 3.25 Calculate the Compton wavelength from the classical electron radius.

In[42]:= $r_e = \dfrac{e^2}{4\pi\ \varepsilon_0\ m\ c^2}$; ScientificForm$\left[\text{UnitConvert}\left[2\pi\ \dfrac{r_e}{\alpha}\right],\ 3\right]$

Out[42]//ScientificForm=

$\quad 2.43 \times 10^{-12}$ m

Thus, we have

$$r_e \sim \alpha\lambda_c \sim \alpha^2 a \sim \frac{\alpha^3}{R_\infty}$$

3.6 RUTHERFORD SCATTERING

Rutherford scattering refers to a hard-scattering process with a r^{-2} force. The original experiment was scattering α particles (helium nucleus) off of a thin metal foil, scattering of nuclei by the electric force at a time before the existence of the nucleus was known. The α particles came from nuclear decay with kinetic energies of about 6 MeV and they are not relativistic.

3.6.1 Effect of the Electrons

Since the α particles are massive, they cannot transfer much energy to an electron. The speed of an α particle with kinetic energy K_α is

$$v_\alpha = \sqrt{\frac{2K_\alpha}{m_\alpha}}.$$

From momentum conservation, the maximum speed of the electron after the collision is

$$v \approx 2v_\alpha = 2\sqrt{\frac{2K_\alpha}{m_\alpha}}.$$

Example 3.26 Calculate the maximum speed of the electron after the collision with an energetic but still non-relativistic α particle.

```
In[43]:= ClearAll["Global`*"];

        Solve[1/2 M V_i^2 == 1/2 M V_f^2 + 1/2 m v^2 && M V_i == M V_f + m v, {v, V_f}]

Out[43]= {{v → 0, V_f → V_i}, {v → (2 M V_i)/(m + M), V_f → (-m V_i + M V_i)/(m + M)}}
```

Example 3.27 In a collision with a 6-MeV α particle with an electron at rest, calculate the maximum kinetic energy of the electron.

```
In[44]:= m = electron PARTICLE [ mass ]; K = 6 MeV; v = 2 √(2K/m_α);

        N[UnitConvert[1/2 m v^2, MeV], 1]

Out[45]= 0.003 MeV
```

This energy is much larger than the energy of outer electrons in atoms (eV scale) but much smaller than the kinetic energy of the α particle. Collisions with electrons are negligible in Rutherford scattering.

3.6.2 Scattering from a Nucleus

The concept of Rutherford scattering is that the scattering angle can be very large if the force is large which can only happen if the α particle comes very close to the nucleus, which can in-turn only happen if the nucleus is tiny on the atomic scale. Suppose the α particle (charge q_α) has an incoming speed

v_α and passes approximately a distance b from the nucleus (charge Q). The momentum transferred (Δp) to the α particle is

$$\Delta p = F\Delta t = \frac{q_\alpha Q}{4\pi\varepsilon_0 b^2}\frac{2b}{v_\alpha},$$

where the time (Δt) for the α partiple to pass the nucleus (the time it spends feeling the large part of the force) is estimated to be $2b/v_\alpha$. The factor of 2 is not important in the estimate; the important part is that the time experiencing the force is proportional to the distance scale divided by the speed. The scattering angle (θ) is

$$\theta = \frac{\Delta p}{p} = \frac{q_\alpha Q}{4\pi\varepsilon_0 b^2}\frac{2b}{v_\alpha}\frac{1}{m_\alpha v_\alpha} = \frac{q_\alpha Q}{4\pi\varepsilon_0 b K_\alpha},$$

where m_α is the α-particle mass and K_α is its kinetic energy.

Example 3.28 Calculate the scattering angle for a 6-MeV α particle that passes a distance 100 fm from a gold nucleus.

In[46]:= `qα = 2 e ; Q = 79 e ; K = 6 MeV ; r = 100 fm ;`

`N[UnitConvert[`$\frac{q_\alpha\,Q}{4\,\pi\,\varepsilon_0\,r\,K}$`], 2]`

Out[47]=

0.38

The above answer is in the dimensionless unit, radians. Even for a relatively large approach distance of 100 fm (the nuclear scale is 1 fm), the scattering angle is quite large, about 22°.

Notice that a 6-MeV α particle cannot penetrate the gold nucleus, which would require a kinetic energy exceeding that of the Coulomb repulsion at a distance equal to the nuclear radius,

$$K_\alpha > \frac{q_\alpha Q}{4\pi\varepsilon_0 r}.$$

Example 3.29 Calculate the α particle kinetic energy needed to penetrate the gold nucleus of radius 7 fm.

In[48]:= `r = 7 fm; N[UnitConvert[`$\frac{q_\alpha\,Q}{4\,\pi\,\varepsilon_0\,r}$`, MeV], 2]`

Out[48]= 33. MeV

3.6.3 Cross Section

Cross section (σ) is a geometrically intuitive quantity that is useful in specifying collision probabilities. It is defined as the transition or scattering rate (R_s), the number of times that something predefined happens per second, divided by the incident flux (Φ),

$$\sigma = \frac{R_s}{\Phi}.$$

Since the numerator has units of s^{-1} and the denominator has units of $m^{-2}s^{-1}$, the unit of cross section is area. Cross section represents the effective geometric size of the target that corresponds to the specified transition.

To get the exact solution for Rutherford scattering, the differential cross section may be written in terms of the so-called impact paramater (b) of Fig. 3.8,

$$d\sigma = 2\pi b db,$$

the area of a ring with radius b and thickness db. Smaller (larger) values of b correspond to larger (smaller) scattering angles. The differential cross section in terms of $x = \cos\theta$ is

$$\frac{d\sigma}{dx} = \frac{d\sigma}{d\cos\theta} = \frac{d\sigma}{db}\frac{db}{d\cos\theta} = 2\pi b\frac{db}{d\cos\theta}.$$

Thus, the problem is reduced to deducing the relationship between impact parameter and scattering angle. This can be done by deducing two expressions for the momentum transfer, one in terms of θ and the other in terms of b.

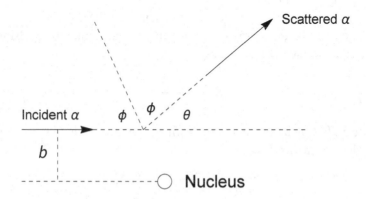

Figure 3.8 An α particle comes within a distance r of a nucleus and scatters at an angle θ. The angle ϕ indicates the axis of symmetry.

Momentum is exchanged in the collision but very little energy (like a ball bouncing off a brick wall) because the nucleus is so massive. Therefore, the direction of the alpha particle changes but not its magnitude and

$$(\Delta p)^2 = p^2 + p^2 - 2p^2 \cos\theta,$$

because the initial and find vectors each of length p and Δp make a triangle. The momentum transfer in the collision is

$$\Delta p = m_\alpha v_\alpha \sqrt{2(1-\cos\theta)}.$$

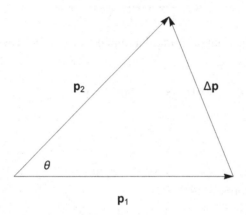

Figure 3.9 The momentum vector before scattering ($\mathbf{p_1}$), after scattering ($\mathbf{p_2}$) and the momentum transfer vector ($\mathbf{\Delta p}$) form a triangle.

A second expression for the momentum transfer comes from integrating the component of force in the direction of momentum transfer,

$$\Delta p = \int dt F \cos\phi = \frac{q_\alpha Q}{4\pi\varepsilon_0} \int dt \frac{\cos\phi}{r^2},$$

and applying conservation of angular momentum (L),

$$L = m_\alpha r^2 \frac{d\phi'}{dt} = m_\alpha v_\alpha b,$$

to get

$$\Delta p = \frac{q_\alpha Q}{4\pi\varepsilon_0 v_\alpha b} \int\limits_{-\phi}^{\phi} d\phi' \cos\phi' = \frac{2q_\alpha Q}{4\pi\varepsilon_0 v_\alpha b} \sin\phi = \frac{2q_\alpha Q}{4\pi\varepsilon_0 v_\alpha b} \sqrt{\frac{1+\cos\theta}{2}}.$$

where the last step comes from the relationship between θ and ϕ,

$$\phi = \frac{\pi}{2} - \frac{\theta}{2}.$$

Example 3.30 Verify the relationship between ϕ and θ.

In[49]:= $Assumptions = 0 < θ < π;

$$\text{Simplify}\left[\text{Sin}\left[\frac{\pi}{2} - \frac{\theta}{2}\right] = \sqrt{\frac{1 + \text{Cos}[\theta]}{2}}\right]$$

Out[49]= True

Equating the two expressions for momentum transfer gives the relationship between impact parameter and scattering angle,

$$b = \frac{q_\alpha Q}{4\pi\varepsilon_0 m_\alpha v_\alpha^2}\sqrt{\frac{1 + \cos\theta}{1 - \cos\theta}}.$$

Example 3.31 Calculate the angular dependance of the cross section, $d\sigma/dx$ where $x = \cos\theta$.

In[50]:= b = $\sqrt{\dfrac{1 + x}{1 - x}}$; b D[b, x] // Simplify

Out[50]= $\dfrac{1}{(-1 + x)^2}$

The whole expression is

$$\frac{d\sigma}{d\cos\theta} = 2\pi b\frac{db}{d\cos\theta} = \frac{\pi}{2}\left(\frac{q_\alpha Q}{4\pi\varepsilon_0 K_\alpha}\right)^2\frac{1}{(1 - \cos\theta)^2}.$$

The cross section for scattering is some specified range comes from integration. This may be easier accomplished using b,

$$\sigma = \int_{b_{\min}}^{b_{\max}} db(2\pi b) = \pi(b_{\max}^2 - b_{\min}^2).$$

Note that $b = 0$ corresponds to $\cos\theta = -1$.

Example 3.32 Calculate the cross section for scattering of a 12 MeV α particle scattering off silver ($Z = 47$) at angles greater than $90°$. Give the answer in barns (1 b = 10^{-28} m).

In[51]:= q_α = 2 e ; Q = 47 e ; K = 12 MeV ; θ = $\dfrac{\pi}{2}$;

$$b = \frac{q_\alpha Q}{4\pi\,\varepsilon_\theta\,(2\,K)}\sqrt{\frac{1 + \text{Cos}[\theta]}{1 - \text{Cos}[\theta]}} ;$$

N[UnitConvert[π b², b], 2]

Out[52]=

   1.0 b

3.6.4 Scattering from a Thin Foil

In performing a scattering experiment, one does not have the liberty to choose the impact parameter. When the α particle passes through a foil, it will encounter nuclei at random values of b. A good way to think of the scattering is that of a flux of nuclei from the target streaming by the α particle. Consider a foil with atomic mass A and density ρ. The number of target nuclei per volume is

$$\frac{N}{V} = \frac{N_0 \rho}{A(10^{-3} \text{ kg})},$$

where N_0 is Avogadro's number.

Example 3.33 Get Avogadro's number.

In[53]:= `Quantity["AvogadroNumber"]`

Out[53]=

N_Θ

The number of target nuclei per area is dN/V where d is the target thickness. If the rate of incoming particles is R_i, then the flux is

$$\Phi = \frac{R_i dN}{V} = \frac{R_i dN_0 \rho}{A(10^{-3} \text{ kg})}.$$

The cross section is

$$\sigma = \frac{R_s}{\Phi} = \frac{R_s A(10^{-3} \text{ kg})}{R_i dN_0 \rho},$$

where R_s is the scattering rate.

Example 3.34 Alpha particles of 6 MeV are incident at normal angle on a 1 μm thick gold foil at a rate of 1000 per s. Calculate the rate that α particles scattered at angles greater than 1 radian.

In[54]:= `q`$_\alpha$` = 2 e ; Q = 79 e ; K = 6 MeV ; θ = .1; b =` $\dfrac{q_\alpha Q}{4 \pi \varepsilon_\Theta (2 K)} \sqrt{\dfrac{1 + \text{Cos}[\theta]}{1 - \text{Cos}[\theta]}}$ `;`

`R =` $\dfrac{1000}{s}$ `; σ = π b`^{2}`; d = 1 microns ; A =` [**gold** ELEMENT][atomic number]`;`

`ρ =` [**gold** ELEMENT][mass density]`;`

`NumberForm`$\Big[$`UnitConvert`$\Big[$ $\dfrac{\sigma R N_\Theta d\rho}{A (10^{-3} \text{ kg})}$ $\Big]$`, 2`$\Big]$

Out[57]//NumberForm=

66. per second

3.7 THE WEAK INTERACTION

3.7.1 Weak Coupling

The strength of the weak interaction is traditionally written in terms of the Fermi constant (G_F). The best value of G_F is given by measurement of the muon lifetime. The muon decay probability goes as G_F^2, so the lifetime goes as the inverse square. The result is

$$G_F = \sqrt{\frac{192\pi^3 \hbar}{(mc^2)^5 \tau}}.$$

Example 3.35 Calculate G_F.

In[58]:= **M =** [**W- boson** PARTICLE] [*mass*] **;**

N $\left[\text{UnitConvert} \left[\alpha \left(\dfrac{1 \text{ GeV}}{\text{M c}^2} \right)^2 \right] , 1 \right]$

Out[59]=

    $1. \times 10^{-6}$

The Fermi constant is stored as Quantity["ReducedFermiConstant"] (G_F^0).

Example 3.36 Get the weak coupling constant.

In[60]:= **Quantity["ReducedFermiConstant"]**

Out[60]=

    G_F^0

The dimension of G_F is inverse energy squared. The dimensionless weak coupling at the energy scale E may be written as

$$\alpha_w = G_F E^2.$$

At 1 GeV,

$$\alpha_w \approx 10^{-5}.$$

3.7.2 Neutrino Cross Section

Event probabilities are measured in particle physics by the interaction cross section (3.6.3). The neutrino has size (cross sectional area) is given by its wave properties as $(\frac{\hbar c}{E})^2$. We may estimate the cross section per nucleon (σ_N) as

$$\sigma_N \approx \alpha_w^2 \left(\frac{\hbar c}{E}\right)^2 .$$

Example 3.37 Estimate the interaction cross section per nucleon for the interaction of a 1 GeV neutrino.

In[61]:= $\mathbf{N}\left[\mathbf{UnitConvert}\left[\left(10^{-6}\right)^2 \pi \left(\dfrac{h\ c}{1\ \text{GeV}}\right)^2\right], 1\right]$

Out[61]=

5. × 10^{-42} m^2

3.7.3 Neutrino Scattering Rate

The cross section that was calculated in 3.7.2 is per nucleon. When calculating the neutrino scattering rate in a target, one must us the same technique derived for Rutherford scattering in a foil 3.6.4.

Example 3.38 Estimate the interaction rate for 1 GeV neutrinos incident on a 10^4 kg target with a flux of $10^8 \text{m}^{-2}\text{s}^{-1}$.

In[62]:= $\mathbf{M} = \mathbf{10^4}\ \text{kg}\ ;\ \Phi = \mathbf{10^8}\ \dfrac{1}{\text{m}^2\ \text{s}}\ ;\ \sigma = \mathbf{5} \times \mathbf{10^{-42}}\ \text{m}^2;$

$\mathbf{N}\left[\mathbf{UnitConvert}\left[\sigma\ \dfrac{M\ N_\theta}{\left(10^{-3}\ \text{kg}\right)}\ \Phi\right], 1\right]$

Out[63]=

0.003 per second

3.7.4 Neutrino Mean Free Path

In passing a distance d through the matter that has n nucleons per volume, a neutrino would encounter nd nucleons per cross-sectional area. For $\sigma_N nd = 1$, one interaction is expected, on the average. The mean free path (9.2.3) is

$$d = \frac{1}{n\sigma_N}.$$

The number of nucleons per volume in the earth is its mass density divided by the mass of a nucleon (0.94 GeV/c^2).

Example 3.39 Estimate the mean free path for a 1 GeV neutrino in the earth.

In[64]:= $n = \dfrac{\boxed{\text{Earth} \;\text{PLANET}} \left[\boxed{\textit{mean density}} \right]}{0.94 \text{ GeV} \Big/ c^2}$; $\sigma_N = 5 \times 10^{-42} \text{ m}^2$;

$\text{NumberForm}\left[\text{UnitConvert}\left[\dfrac{1}{n\,\sigma_N}\right], 1\right]$

Out[65]//NumberForm=

$6. \times 10^{10}$ m

The 1 GeV neutrino mean free path is about 5000 times greater than the earth's diameter (1.3×10^7 m). A 1 GeV neutrino passing through the full diameter of the earth will only have a probability of 1 in 5000 of interacting.

CHAPTER 4

Special Relativity

The first postulate of special relativity states that the laws of physics are identical in all inertial (non-accelerated) frames of reference. The second postulate states that the speed of light in vacuum (c) is the same for all observers.

4.1 BETA AND GAMMA

Calculations in relativity involve the the relative velocity (v) of 2 frames of reference. This is expressed in terms of a dimensionless quantity

$$\beta = \frac{v}{c},$$

which has a range from -1 to 1. Note that β can be negative. Sometimes β is written as a vector,

$$\boldsymbol{\beta} = \frac{\mathbf{v}}{c}.$$

The Lorentz gamma factor (γ) appears frequently:

$$\gamma = \frac{1}{\sqrt{1 - \beta^2}},$$

which has a range of 1 to ∞.

The inverse relationship is

$$\beta = \sqrt{1 - \frac{1}{\gamma^2}}.$$

The γ and β factors are the ingredients of the Lorentz transformation described in 4.5.

As γ gets large, the particle speed is essentially the speed of light. For a γ-factor in the 10^4 range, the difference between particle and light speed is in the m/s range.

DOI: 10.1201/9781003395515-4

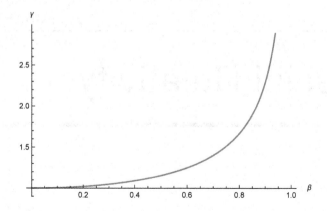

Figure 4.1   As β approaches 1, γ goes rapidly to ∞.

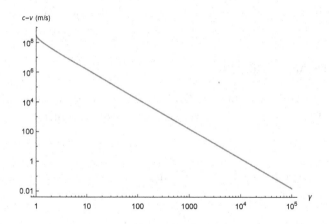

Figure 4.2   The speed of light in vacuum minus particle speed is shown *vs.* γ.

4.2   SPACE AND TIME

4.2.1   Time Dilation

Consider a frame of reference S wherelight moves a distance d from points $\mathcal{A}$ to $\mathcal{B}$ in the y-direction.

Now consider the motion as viewed from a frame S' that moves with a speed v in the $-x$ direction. In this frame both points $\mathcal{A}$ and $\mathcal{B}$ move with speed v in the x direction. In the time that light is traveling, point $\mathcal{B}$ is moving and the light has a longer distance to travel to get there. Its speed is fixed so it takes a longer time to get from $\mathcal{A}$ to $\mathcal{B}$. Since the speed of light is fixed, the two frames are related by the velocity triangle of Fig. 4.4.

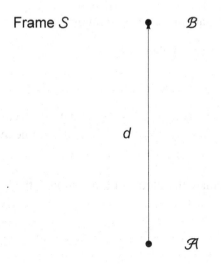

Figure 4.3 In frame $\mathcal{S}$, light moves from $\mathcal{A}$ to $\mathcal{B}$.

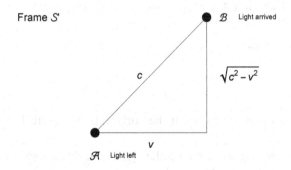

Figure 4.4 In frame $\mathcal{S}'$, in the time that light moves from $\mathcal{A}$ to $\mathcal{B}$, points $\mathcal{A}$ and $\mathcal{B}$ have both moved to the right. The speed of light and the speed v make a right triangle as shown.

In a frame where a clock is moving, time intervals are measured to be longer by a factor of γ longer compared to their measurement in a frame where the clock is at rest,

$$\Delta t' = \gamma \Delta t.$$

An example is the cosmic ray muon that is produced in the upper atmosphere by cosmic ray protons which make energetic short-lived pions which in turn decay into energetic muons. The muons have a lifetime of about 2.2×10^{-6} s in their rest frame. The energetic muons are moving at nearly the speed of light in our frame, having a typical γ of about 10.

Example 4.1 Calculate the speed of a muon that has $\gamma = 10$.

In[1]:= $\beta = \sqrt{1 - \dfrac{1}{\gamma^2}}$; N[UnitConvert[β c /. $\gamma \to$ 10], 3]

Out[1]= 2.98×10^8 m/s

The muon is its own clock ticking away until its decay. The mean lifetime for the muon in its rest frame is about 2.2 μs. In a frame in which it is moving, the lifetime is a factor of γ longer.

Example 4.2 Calculate the lifetime of the muon in the frame where it moves corresponding to $\gamma = 10$.

In[2]:= τ = N[UnitConvert[γ | muon PARTICLE | | mean lifetime |] /. $\gamma \to$ 10, μs], 3]

Out[2]= 22.0 μs

Example 4.3 Calculate the average distance the muon can travel before decaying.

In[3]:= UnitConvert[β c τ /. $\gamma \to$ 10]

Out[3]= 6.55×10^3 m

The muons are observed to reach the surface of the earth because of time dilation.

An atomic clock was flown on the space shuttle at a speed v for a time ΔT as measured on earth. The time interval measured on the space shuttle is shorter by a factor of γ. The difference of the time intervals due to special relativity is

$$\Delta T - \frac{1}{\gamma}\Delta T = T\left(1 - \sqrt{1 - \left(\frac{v}{c}\right)^2}\right).$$

Example 4.4 Calculate the time interval difference for $v = 7710$ m/s and $T = 7$ days.

In[4]:= v = 7710 $\dfrac{m}{s}$; ΔT = 7 days ; N[UnitConvert[$\Delta T \left(1 - \sqrt{1 - \left(\dfrac{v}{c}\right)^2}\right)$, ms], 3]

Out[4]= 0.200 ms

4.2.2 Lorentz Contraction

The concept of length contraction appears together with time dilation. Lengths are shorter when measured in a frame where they are moving. There is one special frame where the length is at rest and in that frame it is the longest.

Consider a stick moving with speed v measured with a stationary clock (frame with no primes). The time for the stick to pass by the clock is L/v. In the frame where the stick is stationary (frame with primes), the clock is moving with speed v. Time intervals are longer in this frame,

$$\Delta t' = \gamma \Delta t = \gamma \frac{L}{v} = \frac{L'}{v},$$

or

$$L = \frac{L'}{\gamma}.$$

Of course, it does not matter in what frame "prime" is used. The physics says the moving clock runs slower and the moving stick is measured to be shorter, as seen in the example of the cosmic-ray muon (4.2).

4.3 ENERGY AND MOMENTUM

Many problems in relativity are more easily visualized in terms of energy-momentum rather than space-time. A moving proton with corresponding β and γ has momentum

$$p = \frac{mv}{\sqrt{1 - \left(\frac{v}{c}\right)^2}} = \gamma \beta mc.$$

The energy is

$$E = \sqrt{(mc^2)^2 + (pc)^2} = \sqrt{(mc^2)^2 + (\gamma \beta mc^2)^2} = \sqrt{(mc^2)^2(1 + \gamma^2 \beta^2)}.$$

Using

$$1 + \gamma^2 \beta^2 = 1 + \frac{\beta^2}{1 - \beta^2} = \frac{1}{1 - \beta^2} = \gamma^2,$$

gives

$$E = \gamma mc^2$$

and

$$\beta = \frac{pc}{E}.$$

In 2022, the CERN Large Hadron Collider (LHC) increased the energy of the protons to 6.8 TeV per beam. One TeV is 10^{12} eV.

Example 4.5 Calculate the gamma factor for LHC protons.

In[5]:= **E = 6.8 TeV ; γ =** $\dfrac{\mathbf{E}}{\boxed{\text{proton PARTICLE}}\left[\boxed{mass}\right] c^2}$

Out[5]= **7247.36**

The gamma factor is so large that β is very nearly equal to 1.

Example 4.6 Calculate $1 - \beta$ for LHC protons.

In[6]:= **ScientificForm**$\left[1 - \sqrt{1 - \dfrac{1}{\gamma^2}}, 2\right]$

Out[6]//ScientificForm=
9.5×10^{-9}

Example 4.7 Calculate how much the speed of LHC protons differ from the speed of light.

In[7]:= **NumberForm**$\left[\text{UnitConvert}\left[\left(1 - \sqrt{1 - \dfrac{1}{\gamma^2}}\right) c\right], 3\right]$

Out[7]//NumberForm=
2.85 m/s

4.4 4-VECTORS

A 4-vector is a 4-component vector with one scalar "time" part and an ordinary vector "space part whose length does not depend on the frame of reference. The length-squared of a 4-vector is defined as the square of the time part minus the square of the space part. The combination $(ct, \mathbf{r})$ makes a 4-vector of length (L) squared:

$$L^2 = (ct)^2 - x^2 + y^2 + z^2).$$

The matrix that flips the sign of the space part,

$$g = \begin{pmatrix} 1 & 0 & 0 & 0 \\ 0 & 1 & 0 & 0 \\ 0 & 0 & 1 & 0 \\ 0 & 0 & 0 & 1 \end{pmatrix},$$

is useful for calculating 4-vector lengths.

Example 4.8 Calculate the length-squared of the space-time 4-vector.

In[8]:= `g = {{1, 0, 0, 0}, {0, -1, 0, 0}, {0, 0, -1, 0}, {0, 0, 0, -1}};`
`X = {ct, x, y, z}; X.(X.g)`

Out[9]= $ct^2 - x^2 - y^2 - z^2$

Other common 4-vectors are energy-momentum $(E, \mathbf{p}c)$, charge-current density $(\rho c, \mathbf{J})$, and electromagnetic scalar-vector potential $(V, c\mathbf{A})$.

4.4.1   Invariant Mass

For any particle, the total energy, momentum, and mass are related by (1.3)

$$(mc^2)^2 = E^2 - (pc)^2,$$

which is precisely the definition of 4-vector length. The addition of 2 4-vectors is also a 4-vector, and for energy-momentum this length is called the invariant mass. For two 4-vectors (E_1, p_1c) and (E_2, p_2c) , the invariant mass squared (s) is

$$s = (E_1 + E_2)^2 - (p_1c + p_2c)^2.$$

4.4.2   Center of Mass

There is a special frame where the two particles have the same magnitude of momentum, in opposite directions. This is commonly referred to as the center of mass (CM) frame. In this case

$$\overset{\bullet}{s} = (E_1^* + E_2^*)^2,$$

where E_1^* and E_2^* are the energies in the CM frame. Thus, the total energy squared in the CM frame is the invariant mass of the system. The collision of two particles is simplest when viewed in the CM frame because the total momentum is zero both before and after the collision. If the two particles have the same mass, then the energies E^* are also equal in the CM frame, and

$$s = (E^* + E^*)^2 = 4E^{*2}.$$

Example 4.9 A proton of momentum $p = 200$ GeV/c strikes another proton at rest. Calculate the energy of each proton in the CM frame.

In[10]:= $p_1 = 200 \dfrac{\text{GeV}}{c}$; $p_2 = 0 \dfrac{\text{GeV}}{c}$; m = [**proton** PARTICLE] [[*mass*]];

$E_1 = \sqrt{(p_1\ c)^2 + (m\ c^2)^2}$;

$E_2 = m\ c^2$;

$N\left[\text{UnitConvert}\left[\dfrac{1}{2}\sqrt{(E_1 + E_2)^2 - (p_1\ c + p_2\ c)^2}\ ,\ \text{GeV}\right]\right]$

Out[11]= 9.70919 GeV

The minimum reaction to make an antiproton in a pp collision is

$$p + p \rightarrow p + p + p + \bar{p}$$

in order to conserve baryon number. The antiproton has the same mass as the proton.

Example 4.10 Calculate the (total) energy that a proton needs to hit another proton at rest and make an antiproton.

In[12]:= `Clear[m, E];`
`Solve`$\left[(E + m\ c^2)^2 - E^2 - (m\ c^2)^2 == (4\ m\ c^2)^2,\ E\right]$

Out[13]=

$\{\{E \to 8\ c^2\ m\}\}$

4.5   LORENTZ TRANSFORMATION

The Lorentz transformation can be written as a 4×4 matrix (Λ) that multiplies a 4-vector. The matrix that corresponds to speed $v = \beta c$ in the x-direction is

$$\Lambda = \begin{pmatrix} \gamma & -\beta\gamma & 0 & 0 \\ -\beta\gamma & \gamma & 0 & 0 \\ 0 & 0 & 1 & 0 \\ 0 & 0 & 0 & 1 \end{pmatrix}.$$

4.5.1   Transformation of Time-Space

Example 4.11 Transform the time-space 4-vector.

In[14]= `ClearAll["Global`*"];`
Λ = `{{`γ`, -`$\beta\,\gamma$`, 0, 0}, {-`$\beta\,\gamma$`, `γ`, 0, 0}, {0, 0, 1, 0}, {0, 0, 0, 1}};`
T = Λ`.{c t, x, y, z}`

Out[15]=

$\{c\,t\,\gamma - x\,\beta\,\gamma,\ x\,\gamma - c\,t\,\beta\,\gamma,\ y,\ z\}$

Example 4.12 Calculate the length-squared of the transformed space-time 4-vector.

```
In[16]:= $Assumptions = {γ > 1, β > 0, β < 1, c > 0, x ∈ ℝ, y ∈ ℝ, z ∈ ℝ, t ∈ ℝ};
        g = {{1, 0, 0, 0}, {0, -1, 0, 0}, {0, 0, -1, 0}, {0, 0, 0, -1}};
                  1
        γ = ─────────── ;
              √(1 - β²)
        T.(T.g) // Simplify
Out[17]= c² t² - x² - y² - z²
```

The Lorentz transformation preserves the length of any 4-vector. The underlying physics states whether something transforms like a 4-vector or not. Time-space makes a 4-vector because the speed of light does not depend on the frame of reference.

The Lorentz transformation accounts for time dilation. and length contraction. Consider the example of the clock. In the frame where the clock is at rest, a time interval $(t_2 - t_1)$ is measured at the same location. Make a Lorentz transformation to the frame where the clock is moving. In this frame the time interval is longer by γ.

Example 4.13 Calculate the time interval $c(t_2 - t_1)$ in the moving frame.

```
In[18]:= (Λ.{c t₂, x, y, z} - Λ.{c t₁, x, y, z})[[1]] // Simplify
Out[18]=
        c (-t₁ + t₂)
        ───────────
          √(1 - β²)
```

4.5.2 Transformation of Energy-Momentum

Energy-momentum makes a 4-vector because the mass of a particle is the same in all frames of reference. The length of energy-momentum 4-vector is

$$\sqrt{E^2 - [(p_xc)^2 + (p_yc)^2 + (p_zc)^2]} = \sqrt{E^2 - (pc)^2} = mc^2.$$

If we transform a mass m from rest, its speed divided by c in the moving frame is seen to be that of the moving frame but in the opposite direction. Its energy is γmc^2 and its momentum is $-\gamma mv\hat{\mathbf{x}}$, where the unit vector is the direction of the transformation.

Example 4.14 Transform a mass m from rest and calculate its β-factor.

In[19]:= $\Lambda = \{\{\gamma, -\beta\gamma, 0, 0\}, \{-\beta\gamma, \gamma, 0, 0\}, \{0, 0, 1, 0\}, \{0, 0, 0, 1\}\};$
$T = \Lambda.\{m c^2, 0, 0, 0\}$
$\frac{T[[2]]}{T[[1]]}$

Out[20]=
$$\left\{ \frac{c^2 m}{\sqrt{1-\beta^2}}, -\frac{c^2 m \beta}{\sqrt{1-\beta^2}}, 0, 0 \right\}$$

Out[21]=
$$-\beta$$

When we transform $(E, \mathbf{p}c)$, the transformed energy is smaller (greater) than the original energy if there is a component of momentum in the same (opposite) direction as the transformation.

Example 4.15 Transform the energy-momentum 4-vector.

In[22]:= $T = \Lambda.\{E, -p c, 0, 0\}$
Out[22]=
$$\left\{ \frac{c p \beta}{\sqrt{1-\beta^2}} + \frac{E}{\sqrt{1-\beta^2}}, -\frac{c p}{\sqrt{1-\beta^2}} - \frac{\beta E}{\sqrt{1-\beta^2}}, 0, 0 \right\}$$

The relative velocity formula can be readily determined from a Lorentz transformation. Consider 2 particles with velocities v and u as shown in Fig. 4.5. The relative speed of the particles is the speed one of the particles in the frame where the other particle is at rest. We can get to this frame with a Lorentz transformation in the x-direction with speed v. The relative speed is then the speed of the moving particle which is obtained by dividing momentum times c by energy.

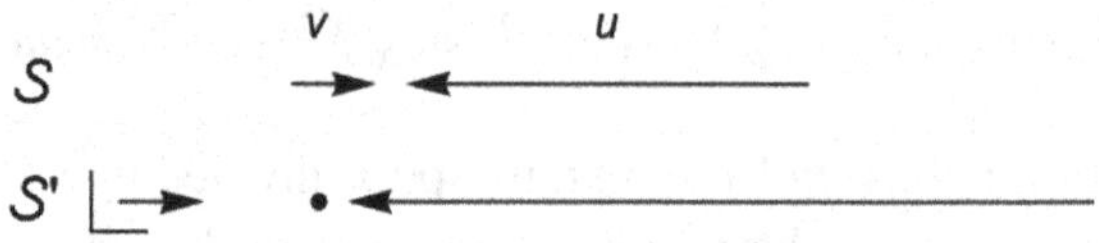

Figure 4.5   In frame S, particles have velocites $v\hat{x}$ and $-u\hat{x}$. In frame S' which moves with speed v in the x-direction, one of the particles is at rest.

Example 4.16 Calculate the relative speeds of the particles in Fig. 4.5.

```
In[23]:= Λ = {{γ, -β γ, 0, 0}, {-β γ, γ, 0, 0}, {0, 0, 1, 0}, {0, 0, 0, 1}};
T = Λ.{E, -p c, 0, 0};
T[[2]]
────── c /. {β → v/c, p → u E/c²} // Simplify
T[[1]]
```

Out[25]=
$$-\frac{c^2\,(u+v)}{c^2+u\,v}$$

4.6 DOPPLER EFFECT

4.6.1 Colinear Light Source

Consider a light source that is moving toward an observer. Let the photon energy be E in the frame where the source is at rest. In the frame where ths source is moving, the photon energy E' is given by the Lorentz trnasformation:

$$E' = \gamma E + \beta\gamma E = E\frac{1+\beta}{\sqrt{1-\beta^2}} = E\sqrt{\frac{1+\beta}{1-\beta}}.$$

Since the frequency is proportional to the energy by $E = hf$ (see 1.4.2), we have

$$f' = f\sqrt{\frac{1+\beta}{1-\beta}},$$

where f' is the frequency in the frame where the source moves and f is the frequency in the frame where the source is at rest. Negative values of β (or switching the sign of β) correspond to the source moving away from the observer.

Example 4.17 The Lyman alpha line (5.4.1) is at $\lambda = 122$ nm for a hydrogen atom at rest. This line is observed to be at $\lambda = 362$ nm when emitted from a moving star. Calculate the speed of the star.

```
In[26]:= $Assumptions = Abs[β] < 1; f/ = c/(362 nm); f = c/(122 nm);

beta = N[Solve[f/ == f √((1+β)/(1-β)), β], 3];

N[UnitConvert[ c β /. beta[[1]]], 3]
```

Out[28]=
$$-2.39 \times 10^8 \text{ m/s}$$

The minus sign indicates the star is moving away.

4.6.2 Redshift

Astrophysicists define redshift (z) to be a fractional change in frequency for stars that are receding from us,

$$z = \frac{f - f'}{f'} = \sqrt{\frac{1 + \beta}{1 - \beta}} - 1$$

Example 4.18 Find β corresponding to redshift z.

In[29]:= $Assumptions = z > 0; Solve$\left[z = \sqrt{\frac{1 + \beta}{1 - \beta}} - 1, \beta\right]$ // Simplify

Out[29]=

$$\left\{\left\{\beta \to \frac{z\,(2 + z)}{2 + 2\,z + z^2}\right\}\right\}$$

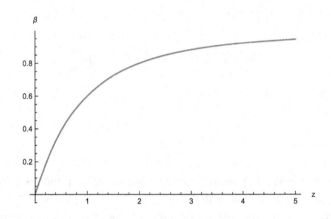

Figure 4.6 Beta *vs.* redshift is plotted for a receding star.

4.6.3 Observation at an Angle

Suppose the light is observed at an angle θ in the lab. Let the photon energy in the lab be E' and its x-component of momentum (along the boost direction) be $E' \cos\theta$. The energy of the photon (E) in its rest frame is given by the Lorentz transformation,

$$E = \gamma E' - \beta\gamma E' \cos\theta,$$

and

$$E' = \frac{E}{\gamma(1 - \beta\cos\theta)}.$$

The same relationship holds for the frequencies,

$$f' = \frac{f}{\gamma(1 - \beta\cos\theta)}.$$

This result is referred to as the "transverse Doppler effect". Note that this result reduces to that of 4.6.1 for approaching source ($\theta = 0$) and receding source ($\theta = \pi$).

The condition of zero redshift is given by $f' = f$. This is the boundary between Doppler redshift and blueshift (increase in frequency)..

Example 4.19 Find $\cos\theta$ for the boundary between redshift and blueshift as a function of β.

In[30]:= $Assumptions = {0 < \beta < 1}$; Solve$\left[1 == \dfrac{\sqrt{1 - \beta^2}}{1 - \beta x}, x\right]$

Out[30]=

$$\left\{\left\{x \to \frac{1 - \sqrt{1 - \beta^2}}{\beta}\right\}\right\}$$

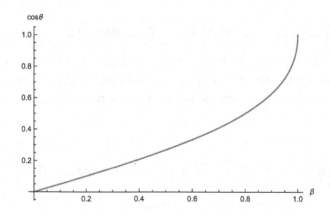

Figure 4.7 Cosine of the boundary angle between redshift and blueshift is plotted as a function of β. Note that at very large values of β there will be a redshift unless the source is headed sharply toward the observer.

CHAPTER 5

Bohr Model

The Bohr model is historically important in the understanding of quantum physics. It gives many important results for the hydrogen atom including atomic size and ionization energy. It is sometimes referred to as a semi-classical calculation, because it includes the physics of angular momentum quantization without the full quantum mechanical description of an electron wave. It is tremendously useful for gaining physical intuition of the quantum nature of the atom.

5.1 QUANTIZATION OF ANGULAR MOMENTUM

In the Bohr model of the hydrogen atom, an electron has a circular orbit with speed v and radius r about the proton. The Bohr quantization condition is that an integer number of electron (de Broglie) wavelengths must fit to the atomic circumference,

$$n\lambda = n\frac{h}{p} = 2\pi r.$$

Written in terms of $\hbar$ with $p = mv$, angular momentum quantization is

$$mvr = n\hbar,$$

where n is a positive integer. Newton's 2nd law gives

$$\frac{e^2}{4\pi\varepsilon_0 r^2} = \frac{mv^2}{r}.$$

The possible values of the orbit radius as a function of the quantum number n are

$$r = \frac{4n^2\pi\hbar^2\varepsilon_0}{e^2 m}.$$

DOI: 10.1201/9781003395515-5

5.2 GROUND STATE

The lowest energy (ground state) corresponds to $n = 1$.

5.2.1 Bohr radius

The value of r corresponding to $n = 1$ is called the Bohr radius (a_0),

$$a_0 = \frac{4\pi\hbar^2\varepsilon_0}{e^2 m}.$$

Example 5.1 Get the Bohr radius.

In[1]:= `Quantity["BohrRadius"]`

Out[1]= a_0

Example 5.2 Compare a_0 with $\frac{4\pi\hbar^2\varepsilon_0}{e^2 m}$.

In[2]:= a_0 == $\dfrac{4\ \pi\ \hbar^2\ \varepsilon_0}{e^2\ \boxed{\textbf{electron}\ \text{PARTICLE}}\left[\boxed{\textit{mass}}\right]}$

Out[2]= `True`

Example 5.3 Calculate the numerical value of the Bohr radius.

In[3]:= `m =` $\boxed{\textbf{electron}\ \text{PARTICLE}}\left[\boxed{\textit{mass}}\right]$ `;`

`N`$\left[\text{UnitConvert}\left[\dfrac{4\ n^2\ \pi\ \hbar^2\ \varepsilon_0}{e^2\ m}, \text{nm}\right] /. n \to 1, 3\right]$

Out[4]= `0.0529 nm`

5.2.2 Energy

The kinetic energy is one-half the potential energy with the opposite sign.

$$K = \frac{1}{2}mv^2 = \frac{1}{2}\frac{e^2}{4\pi\varepsilon_0 r} = -\frac{1}{2}U.$$

The total energy (E) is

$$E = K + U = \frac{V}{2} = -\frac{1}{2}\frac{e^2}{4\pi\varepsilon_0 r}$$

with r given in 5.1 as

$$r = n^2 a_0.$$

Example 5.4 Calculate the total energy for the $n = 1$ orbit.

$$\text{In[5]:= } r = a_0 \text{ ; } N\left[\text{UnitConvert}\left[-\frac{1}{2}\,\frac{e^2}{4\pi\,\varepsilon_0\,r}\,,\,eV\right],\,3\right]$$

$$\text{Out[5]= } -13.6\,eV$$

The electron speed is

$$v = \frac{n\hbar}{mr} = \frac{e^2}{4\pi\varepsilon_0 n\hbar} = \frac{\alpha c}{n}$$

Example 5.5 Calculate the electron speed for $n = 1$ orbit.

$$\text{In[6]:= } N\left[\text{UnitConvert}\left[\alpha\;c\right],\,3\right]$$

$$\text{Out[6]= } 2.19 \times 10^6 \text{ m/s}$$

The electron has an interesting speed. It is extremely fast compared to speeds of macroscopic objects, but yet it is (just barely) nonrelativistic.

Since K is one-half U with the opposite sign, the total energy is $-K$. Therefore, a convenient alternate way to write the total energy is

$$E = -\frac{1}{2}m\left(\frac{\alpha c}{n}\right)^2.$$

5.3   EXCITED STATES

The excited states are given by $n = 2, 3, 4\ldots$

5.3.1   Orbits

The size of the orbits in the Bohr model scale by the square of the quantum number, n^2.

Example 5.6 Calculate the orbit radii for $n = 2$ to 5.

$$\text{In[7]:= } N\left[\text{UnitConvert}\left[\text{Table}\left[n^2\,a_0\,,\,\{n,\,2,\,5\}\right],\,nm\right],\,3\right]$$

$$\text{Out[7]= } \{0.212\,nm\,,\,0.476\,nm\,,\,0.847\,nm\,,\,1.32\,nm\}$$

5.3.2   Speeds

The orbit speeds are inversely proportional to the quantum number n.

Example 5.7 Calculate the orbit speeds for $n = 2$ to 5.

In[8]:= $N\left[\text{UnitConvert}\left[\text{Table}\left[\frac{\alpha\ c}{n}, \{n, 2, 5\}\right], \frac{m}{s}\right], 3\right]$

Out[8]= $\{1.09 \times 10^6 \text{ m/s}, 7.29 \times 10^5 \text{ m/s}, 5.47 \times 10^5 \text{ m/s}, 4.38 \times 10^5 \text{ m/s}\}$

5.3.3 Energies

The energies scale as $-13.6 \text{ eV}/n^2$. Increasing values of the quantum number n give larger energies.

Example 5.8 Calculate the orbit energies for $n = 2$ to 5.

In[9]:= $N\left[\text{UnitConvert}\left[\text{Table}\left[-\frac{1}{2}\frac{e^2}{4\pi\ \varepsilon_0\ n^2\ a_0}, \{n, 2, 5\}\right], \text{eV}\right], 3\right]$

Out[9]= $\{-3.40 \text{ eV}, -1.51 \text{ eV}, -0.850 \text{ eV}, -0.544 \text{ eV}\}$

5.4 TRANSITIONS BETWEEN ENERGY LEVELS

When an electron moves from a higher energy orbit to a lower energy orbit, a photon is emitted with an energy equal to the energy difference between the orbits.

5.4.1 Lyman Series

Transistons to the $n = 1$ (ground) state are called the Lyman series. In Ex. 5.9, a function is defined (A.9) to evaluate the energy $E(n)$. A trick is played to avoid use of the reserved E (A.4) by writing it as a Greek letter which looks nearly identical in Mathematica.

Example 5.9 Get the first 4 Lyman series energies.

In[10]:= $Ε[n\_] = -\frac{1}{2}\frac{e^2}{4\pi\ \varepsilon_0\ n^2\ r};$

$\text{Table}\left[N\left[\text{UnitConvert}\left[Ε[n] - Ε[1], \text{eV}\right], 3\right], \{n, 2, 5\}\right]$

Out[11]=
$\{10.2 \text{ eV}, 12.1 \text{ eV}, 12.8 \text{ eV}, 13.1 \text{ eV}\}$

These photons are in the ultraviolet region.

Example 5.10 Get the first 4 Lyman series wavelengths.

In[12]:= $\text{Table}\left[\text{N}\left[\text{UnitConvert}\left[\dfrac{h\ c}{E[n]-E[1]},\ nm\right],\ 3\right],\ \{n,\ 2,\ 5\}\right]$

Out[12]=

$\{\,122.\ nm\ ,\ 103.\ nm\ ,\ 97.2\ nm\ ,\ 94.9\ nm\,\}$

5.4.2 Balmer Series

Transitions to the $n = 2$ (1st excited) state are called the Balmer series.

Example 5.11 Get the first 4 Balmer series energies.

In[13]:= $\text{Table}\left[\text{N}\left[\text{UnitConvert}\left[E[n]-E[2],\ eV\,\right],\ 3\right],\ \{n,\ 3,\ 6\}\right]$

Out[13]=

$\{\,1.89\ eV\ ,\ 2.55\ eV\ ,\ 2.86\ eV\ ,\ 3.02\ eV\,\}$

These photons are in the visible region.

Example 5.12 Get the first 4 Balmer series wavelengths.

In[14]:= $\text{Table}\left[\text{N}\left[\text{UnitConvert}\left[\dfrac{h\ c}{E[n]-E[2]},\ nm\right],\ 3\right],\ \{n,\ 3,\ 6\}\right]$

Out[14]=

$\{\,656.\ nm\ ,\ 486.\ nm\ ,\ 434.\ nm\ ,\ 410.\ nm\,\}$

5.4.3 Paschen Series

Transitions to the $n = 3$ (2nd excited) state are called the Paschen series.

Example 5.13 Get the first 4 Paschen series energies.

In[15]:= $\text{Table}\left[\text{N}\left[\text{UnitConvert}\left[E[n]-E[3],\ eV\,\right],\ 3\right],\ \{n,\ 4,\ 7\}\right]$

Out[15]= $\{0.661\ eV,\ 0.968\ eV,\ 1.13\ eV,\ 1.23\ eV\}$

These photons are in the infrared region.

Example 5.14 Get the first 4 Paschen series wavelengths.

In[16]:= $\text{Table}\left[\text{N}\left[\text{UnitConvert}\left[\dfrac{h\ c}{E[n]-E[3]},\ nm\right],\ 3\right],\ \{n,\ 4,\ 7\}\right]$

Out[16]=

$\{\,1.87\times 10^{3}\ nm\ ,\ 1.28\times 10^{3}\ nm\ ,\ 1.09\times 10^{3}\ nm\ ,\ 1.00\times 10^{3}\ nm\,\}$

5.5   RYDBERG CONSTANT

The Rydberg constant (R_∞) is defined to be the inverse wavelength of a photon corresponding to a transiton from $n = \infty \to n = 1$.

$$R_\infty = -\frac{E_1}{hc}.$$

Example 5.15 Calculate the Rydberg constant.

In[17]:= **UnitConvert$\left[\dfrac{-E[1]}{h\ c}\right]$**

Out[17]=

1.097373157×10^7 per meter

Example 5.16 Get the Rydberg constant.

In[18]:= **Quantity["RydbergConstant"]**

Out[18]= R_∞

In terms of R_∞, the wavelength of a photon from the transition $j \to i$ is given by

$$\frac{1}{\lambda} = R_\infty \left(\frac{1}{n_i^2} - \frac{1}{n_j^2}\right).$$

Example 5.17 Calculate the energy of the first Lyman transition using the Rydberg constant.

In[19]:= **N$\left[\text{UnitConvert}\left[h\ c\ R_\infty\ \left(\dfrac{1}{n_1{}^2} - \dfrac{1}{n_2{}^2}\right) / . \{n_1 \to 1, n_2 \to 2\},\ eV\right], 3\right]$**

Out[19]=

10.2 eV

5.6   REDUCED MASS

The Bohr model assumes that the proton is stationary. We can correct for this by working in the center-of-mass. The electron mass is then replaced by

$$\mu = \frac{m m_p}{m + m_p}.$$

Example 5.18 Calculate the reduced mass divided by the electron mass.

In[20]:= **m** = [**electron** PARTICLE] [*mass*] ; **M** = [**proton** PARTICLE] [*mass*] ;

$$\mu = \frac{m\,M}{m + M} \; ; \; \frac{\mu}{m}$$

Out[21]=

0.999455679

The deuteron has a slightly bigger reduced mass than the proton. Consider the Balmer series $n = 3$ to $n = 2$ transition. This photon from this transition was calculated to have a wavelength of 656 nm (5.4.2). The photon energy is

$$E = \frac{1}{2}\mu(\alpha c)^2 \left(\frac{1}{4} - \frac{1}{9}\right),$$

and the wavelength is

$$\lambda = \frac{hc}{E}.$$

Example 5.19 Calculate the wavelength difference between the $n = 3$ to $n = 2$ transition in hydrogen *vs.* deuterium.

In[22]:= **M_d** = [**deuteron** PARTICLE] [*mass*] ; $\mu_d = \frac{m\,M_d}{m + M_d}$;

N[UnitConvert[$\frac{h\ c}{\frac{1}{2}\,\mu\,(\alpha\ c\,)^2\,\left(\frac{1}{4} - \frac{1}{9}\right)} - \frac{h\ c}{\frac{1}{2}\,\mu_d\,(\alpha\ c\,)^2\,\left(\frac{1}{4} - \frac{1}{9}\right)}$, nm], 2]

Out[23]=

0.18 nm

5.7 COLLAPSE OF THE BOHR ATOM

The power (P) radiated by a charged particle is proportional to the square of the acceleration (a). For a non-relativistic charge e, the result is

$$P = \frac{e^2 a^2}{6\pi\varepsilon_0 c^3}.$$

For an atom in a circular orbit, the acceleration is

$$a = \frac{v^2}{r}.$$

Example 5.20 Calculate the power radiated for a classical electron in the Bohr orbit.

In[24]:= $r = a_\theta ; a = \dfrac{(\alpha\ c)^2}{r};$

$N\left[UnitConvert\left[\dfrac{e^2 a^2}{6\pi\ \varepsilon_0\ c^3}, \dfrac{eV}{s}\right], 1\right]$

Out[25]=

$3. \times 10^{11}\ eV/s$

The energy radiated per second is enormous compared to the kinetic energy of the electron and the Bohr atom would collapse on a time scale of 10^{-11} s.

5.8   CORRESPONDENCE PRINCIPLE

5.8.1   Orbit and Radiation Frequency

In the Bohr model, the orbit frequency from which an electron radiates when moving to the next lower state,

$$f = \frac{\alpha c/n}{2\pi n^2 a_0},$$

is smaller than the frequency of the emitted photon

$$f = \frac{E_{n+1} - E_n}{h}.$$

For large values of the quantum number n, we must get agreement with classical physics in which the orbit frequency (f) of the charge gives radiation of the same frequency.

Example 5.21 Calculate the orbit frequency.

In[26]:= $Clear[v, m] ; r = \dfrac{4\ n^2\ \pi\ \hbar^2\ \varepsilon_\theta}{e^2\ m} ; v = \dfrac{n\ \hbar}{m\ r} ; f = \dfrac{v}{2\pi r}$

Out[26]=

$\dfrac{e^4\ m}{32\ n^3\ \pi^3\ \hbar^3\ \varepsilon_\theta^2}$

Example 5.22 Calculate the radiation frequency and take the limit as $n \to \infty$.

In[27]:= $\text{Series}\left[\dfrac{E[n+1] - E[n]}{2\pi\hbar}, \{n, \infty, 3\}\right] \text{ // Simplify}$

Out[27]=

$$\dfrac{\frac{1}{8\,n^2}\,e^2/\,(a_0\,\varepsilon_0)}{\hbar\,n^3} + O\left[\dfrac{1}{n}\right]^4$$

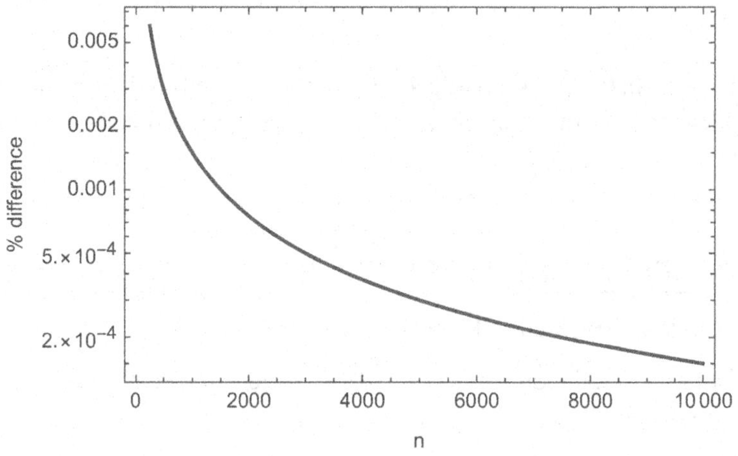

Figure 5.1 The percentage difference between the radiation and orbit frequencies drops *vs.* the quantum number n.

5.8.2 Earth's Orbit

We may apply the Bohr model to the earth's orbit about the sun by replacing the electric force strength ($\frac{e^2}{4\pi\varepsilon_0}$) with the gravitational force strength ($Gm_e M_s$) in the expression for r calculated in 5.1, giving

$$r = \dfrac{n^2\hbar^2}{Gm_e^2 M_s}.$$

Example 5.23 Calculate the quantum number for the earth's orbit about the sun.

In[28]:=

m = [**Earth** PLANET][*mass*]; M = [**Sun** STAR][*mass*];

r = [**Earth** PLANET][*orbital semimajor axis*];

$N\left[\text{Sqrt}\left[\text{UnitConvert}\left[\dfrac{G\ r\,m^2\ M}{\hbar^2}\right]\right], 1\right]$

Out[29]=

$3. \times 10^{74}$

5.8.3 LHC Proton

Protons orbit the Large Hadron Collider (LHC), which has a circumference of 27 km, with a momentum of 13.6 TeV/c. Quantization of angular momentum gives

$$pr = n\hbar.$$

Example 5.24 Calculate the quantum number for LHC protons.

In[30]:= $\mathbf{p} = 13.6 \dfrac{\text{TeV}}{\text{c}}$; $\mathbf{r} = \dfrac{27 \text{ km}}{2\pi}$; NumberForm$\left[\dfrac{\text{p r}}{\hbar}, 1\right]$

Out[30]//NumberForm=

$3. \times 10^{23}$

CHAPTER 6

Particle in a Box

The particle in a box, also referred to as "infinite square well" is an important example because it illustrates the many features of the solution to the Schrödinger equation in an example requiring minimal math. The problem will be worked first in one dimension.

6.1   THE POTENTIAL

The box is defined by a potential that is zero in some region $(0 < x < L)$ and infinite at the boundaries. Thus, the particle is absolutely confined and the wave function is zero beyond the boundaries.

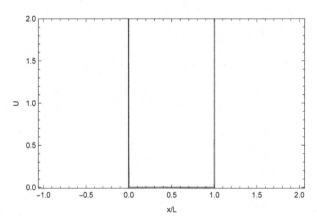

Figure 6.1   The potential for the one-dimensional particle in a box becomes infinite at the boundaries.

DOI: 10.1201/9781003395515-6

6.2   THE SCHRÖDINGER EQUATION

The Schrödinger equation inside the box is

$$\frac{d^2\psi}{dx^2} = -\frac{2mE}{\hbar^2}\psi.$$

This is the most important differential equation that appears in physics. It is of the same form that describes mechanical oscillations: two derivatives of a function gives the same function back with a negative sign and a constant multiplier. The only functions that satisfy this are the sine and cosine. The sine and cosine have the same shape and differ only by placement of the origin.

6.3   SOLUTION

There are multiple solutions (in this case, an infinite number) corresponding to the number of oscillations that fit inside the box. The general solution that satisfies the boundary conditions ($\psi \to 0$. at at $x = 0$ and $x = L$) is

$$\psi = A\sin\left(\frac{n\pi x}{L}\right),$$

where A is a constant that is determined by normalization to unit probability and n is a positive integer that tells us how many peaks the sine function makes inside the box. It is a general feature in the quantum world that the lowest energy cannot be zero because that would require $\psi = 0$ which would correspond to no particle at all.

Example 6.1 Solve the Schrödinger equation for a particle in a box.

In[1]:= ψ = A Sin$\left[\frac{n\pi x}{L}\right]$; Solve$\left[\partial_x\partial_x\psi = -\frac{(2 m E)\psi}{\hbar^2}, E, \mathbb{R}\right]$.{1}

Out[1]= $\left\{E \to \frac{n^2\pi^2\hbar^2}{2 L^2 m}\right\}$

The energy levels increase with n, but the percentage of increase decreases with n, in such a manner that we arrive at the classical limit at large n.

The interpretation of the wave function is that its square is the probability per distance x (or per volume for 3 dimensions) of finding the particle at location x. The integral of the wave function squared over all possible positions

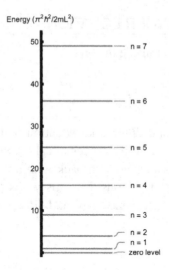

Figure 6.2 The energy levels are indicated in units of $\frac{\pi^2 \hbar^2}{2mL^2}$.

is normalized to unity.

Example 6.2 Find the normalization constant A.

In[2]:= $Assumptions = {L > 0, A > 0, n ∈ Integers};

Solve$\left[A^2 \int_0^L Sin\left[\frac{n \pi x}{L}\right]^2 dx == 1, A, Reals\right]$.{1} // Simplify

Out[2]= $\left\{A \to \frac{\sqrt{2}}{\sqrt{L}}\right\}$

6.3.1   Wave Functions

The wave functions for the 3 lowest energy states are shown in fig. 6.3. The square of the wave functions is shown in fig. 6.4.

6.3.2   Electron in a Box

For an electron in a box of the atomic size (0.2 nm), we expect to get a ground state energy on the eV scale.

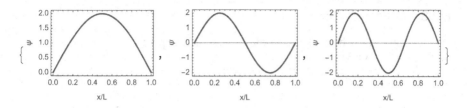

Figure 6.3   Particle-in-a-box wave functions for $n = 1, 2,$ and 3.

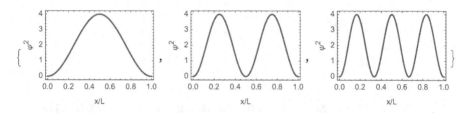

Figure 6.4   Particle-in-a-box wave functions squared for $n = 1, 2,$ and 3.

Example 6.3 Calculate the ground state energy for an electron in a box of width 0.2 nm.

In[3]:= **L = 0.2 nm ; m =** [**electron** PARTICLE] [*mass*] **; n = 1;**

$$\text{NumberForm}\left[\text{UnitConvert}\left[\frac{n^2 \, \pi^2 \, \hbar^2}{2 \, L^2 \, m}, \text{ eV}\right], 2\right]$$

Out[4]//NumberForm=

 9.4 eV

We may make a comparison with the Bohr model (Chap. 5) which requires

$$\lambda = 2\pi a_0.$$

This is the $n = 2$ state for a box of size $L = 2\pi a_0$.

Example 6.4 Calculate the $n = 2$ energy for an electron in a box of width $L = 2\pi a_0$.

In[5]:= **L = 2 π a$_0$; n = 2; N** $\left[\text{UnitConvert}\left[\dfrac{n^2 \, \pi^2 \, \hbar^2}{2 \, L^2 \, m}, \text{ eV}\right], 3\right]$

Out[5]=   13.6 eV

6.3.3 Proton in a Box

For a proton in a box of the nucleus size, we expect to get a ground state energy on the MeV scale.

Example 6.5 Calculate the energy for a proton in a box of width 3 fm.

In[6]:= `L = 3 fm ; n = 1; m = [ proton PARTICLE ] [ mass ];`

$$\text{NumberForm}\left[\text{UnitConvert}\left[\frac{n^2\,\pi^2\,\hbar^2}{2\,L^2\,m}, \text{MeV}\right], 2\right]$$

Out[7]//NumberForm=

23. MeV

6.4 COMPARISON WITH THE DE BROGLIE WAVELENGTH AND THE UNCERTAINTY PRINCIPLE

The minimum energy from the particle in a box calculation is

$$\Delta E > \frac{\pi^2 \hbar^2}{2mL^2} = \frac{h^2}{8mL^2}.$$

In the de Broglie picture, this can be thought of as a standing wave where 1/2 wavelength fits inside the box,

$$\frac{\lambda}{2} = L.$$

Thus gives

$$\Delta p = \frac{h}{\lambda} = \frac{h}{2L},$$

corresponding to a minimum kinetic energy,

$$\Delta E > \frac{(\Delta p)^2}{2m} = \frac{h^2}{8mL^2}.$$

The de Broglie wavelength interpretation gives the same minimum energy as the Schrödinger equation.

To apply the uncertainty principle, the uncertainty Δx must be evaluated. This may be taken to be the rms of a flat distribution.

Example 6.6 Calculate the rms position if a flat probability from 0 to L.

In[8]:= `Clear[L]; $Assumptions = L > 0;`

$$\sqrt{\frac{1}{L} \text{Integrate}\left[\left(x - \frac{L}{2}\right)^2, \{x, 0, L\}\right]} \text{ // Simplify}$$

Out[9]= $\dfrac{L}{2\sqrt{3}}$

For

$$\Delta x = \frac{L}{2\sqrt{3}},$$

the uncertainty principle gives

$$\Delta p = \frac{\hbar}{2\Delta x} = \frac{h\sqrt{3}}{2\pi L}.$$

The minimum allowed kinetic energy

$$\Delta E > \frac{(\Delta p)^2}{2m} = \left(\frac{3}{\pi^2}\right)\frac{h^2}{8mL^2}.$$

In this case, the minimum kinetic energy is greater than required by the uncertainty principle by about a factor of 3. To meet the uncertainty principle minimum, a Gaussian distribution is required.

6.5 EXPECTATION VALUES

6.5.1 Position

The wave function squared is the probability distribution and its average value is referred to as the "expectation value". The probability distributions look very different for different values of n but they all have an average value of $L/2$. There is an equal probability of the particle being found on the left- or right-hand half of the box. The wave function squared is symmetric about $x = L/2$.

Example 6.7 Calculate the average value of x for $n = 1, 2,$ and 3.

In[9]:= `ClearAll["Global`*"];`

In[10]:= $\text{Table}\left[\psi = \sqrt{\frac{2}{L}} \text{Sin}\left[\frac{n \pi x}{L}\right]; \int_0^L \psi^2 x \, dx \, /. \, n \to i, \{i, 1, 3\}\right]$

Out[10]=

$$\left\{\frac{L}{2}, \frac{L}{2}, \frac{L}{2}\right\}$$

Example 6.8 Calculate the average value of x^2 for $n = 1, 2,$ and 3.

In[11]:= Table$\left[\int_{0}^{L}\psi^2\, x^2\, dx\ /.\ n\to i,\ \{i,\ 1,\ 3\}\right]$ // Simplify

Out[11]=

$$\left\{\frac{1}{6}\, L^2\left(2 - \frac{3}{\pi^2}\right),\ \frac{1}{6}\, L^2\left(2 - \frac{3}{4\,\pi^2}\right),\ \frac{1}{6}\, L^2\left(2 - \frac{1}{3\,\pi^2}\right)\right\}$$

6.5.2 Momentum

The momentum operator is given by

$$p = -i\hbar\frac{d}{dx}.$$

Example 6.9 Calculate the average value of momentum for $n = 1, 2,$ and 3.

In[12]:= Table$\left[\int_{0}^{L}\psi\,(-i\,\hbar\,\partial_x\psi)\, dx\ /.\ n\to i,\ \{i,\ 1,\ 3\}\right]$

Out[12]=

$\{0,\ 0,\ 0\}$

Example 6.10 Calculate the average value of momentum squared for $n = 1,$ 2, and 3.

In[13]:= Table$\left[\int_{0}^{L}\psi\,(-i\,\hbar\,\partial_x\,(-i\,\hbar\,\partial_x\,\psi))\, dx\ /.\ n\to i,\ \{i,\ 1,\ 3\}\right]$

Out[13]= $\left\{\dfrac{\pi^2\,\hbar^2}{L^2},\ \dfrac{\pi^2\,\hbar^2}{L^2},\ \dfrac{\pi^2\,\hbar^2}{L^2}\right\}$

The average value of momentum squared is seen to give the energy,

$$E = \frac{(p^2)_{\text{ave}}}{2m}.$$

6.6 CONSISTENCY WITH THE UNCERTAINTY PRINCIPLE

We may estimate the uncertainty in the knowledge of the position and momentum from the root-mean-square values which have been calculated in 6.5.

Example 6.11 Check that the product of the root-mean-square values of position and momentum satisfy the uncertainty principle.

In[14]:= $Assumptions = L > 0; Δx = $\sqrt{\dfrac{L^2\left(-3+2\pi^2\right)}{6\pi^2}}$ // Simplify; Δp = $\dfrac{\pi \hbar}{L}$;

N[Δx Δp, 2]

Out[15]=

1.7 $\hbar$

This value is larger than $\hbar/2$ as required by the uncertainty principle.

6.7   CORRESPONDENCE PRINCIPLE

When the quantum number n is large, the energy levels become close together and we get so many wiggles in the wave function that we get agreement with classical physics.

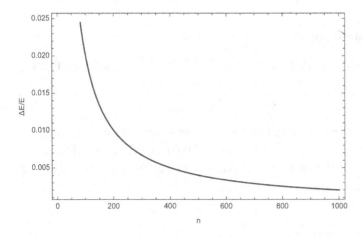

Figure 6.5 The fractional energy difference between adjacent levels is indicated *vs. n*.

6.7.1   Phone in Box

For a macroscopic object we expect the ground state energy will be zero and also that any measurable energy corresponds to a huge quantum number by the correspondence principle. The iPhone 14 has a mass of 6.07 oz (ounce is a mass unit in Mathematica).

Example 6.12 Calculate the ground state energy of a mobile phone confined to be in a room of width 15 feet.

In[16]:= `L = 15 ft ; m = 6.07 oz ; n = 1; E = UnitConvert[` $\dfrac{n^2 \pi^2 \hbar^2}{2 L^2 m}$ `, J];`

`NumberForm[%, 1]`

Out[17]//NumberForm=

$2. \times 10^{-68}$ J

Example 6.13 Calculate the quantum number n if the iPhone had a kinetic energy of a thermal photon (1/40 eV).

In[18]:= `NumberForm[UnitConvert[` $\sqrt{\dfrac{(1/40)\ eV}{E}}$ `], 1]`

Out[18]//NumberForm=

$5. \times 10^{23}$

6.7.2 Classical Probability

Consider the probability that the particle is at the center of the box in the tiny interval

$$\frac{L}{2} - \frac{L}{200} < x < \frac{L}{2} + \frac{L}{200}$$

The classical answer is 0.01 independent of the size of the box because the interval has been chosen to be a fraction (1/100) of the length. The $n = 1$ wave function has a much larger probability of being found near there, in fact, nearly twice as large.

Example 6.14 Integrate ψ^2 from $\frac{L}{2} - \frac{L}{200} < x < \frac{L}{2} + \frac{L}{200}$.

In[19]:= `ClearAll["Global`*"]; ` $\psi = \sqrt{\dfrac{2}{L}}\ \mathrm{Sin}\left[\dfrac{n\pi x}{L}\right];$

`int =` $\displaystyle\int_{(L/2)-(L/200)}^{(L/2)+(L/200)} \psi^2\, dx$

Out[20]=

$$\dfrac{n\pi + 50\,\mathrm{Sin}\left[\frac{99\,n\pi}{100}\right] - 50\,\mathrm{Sin}\left[\frac{101\,n\pi}{100}\right]}{100\,n\pi}$$

The $n = 2$ wave function has a node there and has nearly zero probability of being found near there.

Example 6.15 Calculate the probability that the particle is found at the center of the box within 1% for $n = 1$ and 2.

In[21]:= `{N[int /. n → 1, 3], N[int /. n → 2, 3]}`

Out[21]=
$$\{0.0200, 6.58 \times 10^{-6}\}$$

The correspondence principle tells us that at large n, when we get enough wiggles, the probability will go to the classical value of 0.01.

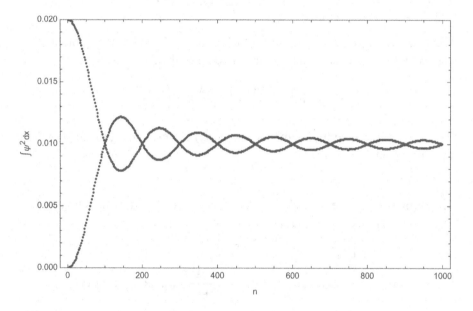

Figure 6.6 The probability that the particle is found at the center of the box within 1% is shown for $n = 1$ to 1000.

The over and undershoot gets progressively smaller with increasing n. From symmetry of the wave function, we always get the exact answer when n is a multiple of 100 and the worst case when n is 50 plus a multiple of 100. How close is the solution to the classical result when $n = 10^6 + 50$?

Example 6.16 Calculate the probability that the particle is found at the center of the box within 1% for $n = 10^6 + 50$.

```
In[22]:=  N[int /. n → 10^6 + 50]
```
Out[22]=

   0.00999968

6.8 THREE DIMENSIONS

For three dimensions, the wave function is $\psi(x, y, x)$. The Schrödinger equation inside the box is

$$\frac{d^2\psi}{dx^2} + \frac{d^2\psi}{dy^2} + \frac{d^2\psi}{dz^2} = -\frac{2mE}{\hbar^2}\psi.$$

Choosing the size of the box to be $L_1 \times L_2 \times L_3$, the solution is obtained by the technique of separation of variables to be of the form

$$\psi = A \sin\left(\frac{n_1\pi x}{L_1}\right)\sin\left(\frac{n_2\pi x}{L_2}\right)\sin\left(\frac{n_3\pi x}{L_3}\right),$$

There are now 3 quantum numbers (n_1, n_2, n_3) corresponding to the boundary conditions on x, y, and z.

Example 6.17 Solve the Schrödinger equation for a particle in a box in 3D.

```
In[23]:=  ClearAll["Global`*"];
          ψ = A Sin[ n₁ π x / L₁ ] Sin[ n₂ π y / L₂ ] Sin[ n₃ π z / L₃ ];
          Solve[∂ₓ∂ₓψ + ∂ᵧ∂ᵧψ + ∂_z∂_zψ == - (2 m E) ψ / ħ², E, ℝ] // Simplify
```
Out[23]=

$$\left\{\left\{E \rightarrow \frac{\pi^2\hbar^2\left(\frac{n_1^2}{L_1^2} + \frac{n_2^2}{L_2^2} + \frac{n_3^2}{L_3^2}\right)}{2\,m}\right\}\right\}$$

The ground-state energy is obtained when by

$$n_1 = n_2 = n_3 = 1.$$

The ordering of the energies after that will depend on the relative sizes of $L_1, L_2,$ and L_3. The energy levels can be degenerate. The greatest degeneracy will happen when $L_1 = L_2 = L_3$. In that case (n_1, n_2, n_3) being $(1,1,2)$, $(1,2,1)$, or $(2,1,1)$ all have the same energy.

6.9 FINITE POTENTIAL

For a box where the sides are not infinitely high ("finite square well") the Schrödinger equation becomes

$$\frac{d^2\psi}{dx^2} = \frac{2m(V-E)}{\hbar^2}\psi,$$

where V is the value of the potential outside the box. The principle difference in the solution, compared to the infinite potential, is that now the wave function extends outside the box. It must go to 0 at $\pm\infty$ instead of at the boundary of the box.

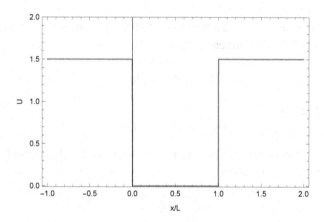

Figure 6.7   The potential for a finite square well is constant outside the boundary.

6.9.1   Solution Technique

Defining

$$\alpha = \sqrt{\frac{2m(V-E)}{\hbar^2}},$$

the solution for $x < 0$ becomes

$$\psi = Ae^{\alpha x},$$

and for $x > L$,

$$\psi = Be^{-\alpha x}.$$

Inside the box, the Schrödinger equation is identical to the infinite potential.

$$\frac{d^2\psi}{dx^2} = -\frac{2mE}{\hbar^2}\psi.$$

Defining

$$k = \sqrt{\frac{2mE}{\hbar^2}},$$

the solution may be written

$$\psi = Ce^{ikx} + De^{-ikx}.$$

The remaining task is to match the boundary conditions at $x = 0$ and $x = L$. Continuity of ψ at $x = 0$ gives

$$A = C + D,$$

and at $x = L$ gives

$$Be^{-\alpha L} = Ce^{ikL} + De^{-ikL}.$$

The derivative $\frac{d\psi}{dx}$ must also be continuous in order for the second derivative to exist. This gives at $x = 0$,

$$\alpha A = ikC - ikD,$$

and at $x = L$

$$-\alpha Be^{-\alpha L} = ikCe^{ikL} - ikDe^{-ikL}.$$

6.9.2 Energy Condition

The relationship between α and k gives the allowed energies. It is useful to separate these into 2 classes according to the symmetry of the wave function. The even functions have $\psi(x) = \psi(-x)$. The ground state will be even. The excited states, if they exist, will alternate between even and odd, $\psi(x) = -\psi(-x)$.

The even states are obtained eliminating the constants making use of $\psi(x) = \psi(-x)$. The conditions at $x = 0$ give

$$C = \frac{-\alpha - ik}{\alpha - ik}D,$$

and combining this with the condition at $x = L$ gives the relationship between α and k to be

$$-\alpha \frac{-\alpha - ik}{\alpha - ik} e^{ikL} - \alpha e^{-ikL} = ik \frac{-\alpha - ik}{\alpha - ik} e^{ikL} - ike^{-ikL}.$$

The even functions have $kL < \pi$ and the wave function for $0 < x < L$ has the form

$$\psi \sim \cos\left[k\left(x - \frac{L}{2}\right)\right].$$

Example 6.18 Solve for the relationship between α and k for the even functions.

```
In[24]:= $Assumptions = {α > 0, k > 0, L > 0, k L < π};

        Solve[-α -α - i k/α - i k e^i k L - α e^-i k L == i k -α - i k/α - i k e^i k L - i k e^-i k L,

        {α}].{1} // FullSimplify

Out[24]= {α → k Tan[k L/2]}
```

The equation for the energy is

$$\sqrt{\frac{2m(V - E)}{\hbar^2}} = \sqrt{\frac{2mE}{\hbar^2}} \tan\left(\sqrt{\frac{2mE}{\hbar^2}} \frac{L}{2}\right).$$

This is a transcendental equation which must be solved numerically.

The odd functions have $\pi < kL < 2\pi$ and the wave function for $0 < x < L$ has the form

$$\psi \sim \sin\left[k\left(x - \frac{L}{2}\right)\right].$$

Example 6.19 Solve for the relationship between α and k for the odd functions.

```
In[25]:= $Assumptions = {α > 0, k > 0, L > 0, π < k L < 2 π};

        Solve[-α -α - i k/α - i k e^i k L - α e^-i k L == i k -α - i k/α - i k e^i k L - i k e^-i k L,

        {α}].{1} // FullSimplify

Out[25]= {α → -k Cot[k L/2]}
```

The equation for the energy is

$$\sqrt{\frac{2m(V - E)}{\hbar^2}} = -\sqrt{\frac{2mE}{\hbar^2}} \cot\left(\sqrt{\frac{2mE}{\hbar^2}} \frac{L}{2}\right).$$

This is another transcendental equation which must be solved numerically.

6.9.3 Solving for the Energy

Define the dimensionless variables

$$W = \frac{mVL^2}{2\hbar^2}$$

and

$$\xi = \frac{L\sqrt{2mE}}{2\hbar}$$

to write the transcendental equation in dimensionless form for the even states as

$$\tan\xi = \frac{\sqrt{W - \xi^2}}{\xi},$$

and the odd states as

$$\tan\xi = -\frac{\xi}{\sqrt{W - \xi^2}}.$$

Given numerical values of V and L we can solve for the allowed energies. The dimensionless transcendental equation allows visualization of the allowed solutions graphically. There is always at least 1 bound state but there could be any number in general.

Consider an electron with $V = 100$ eV and $L = 0.2$ nm. Figure 6.8 shows that there are two solutions for the energy where $\tan\xi$ and $\frac{\sqrt{W - \xi^2}}{\xi}$ are equal.

Example 6.20 Find the energies of the even states.

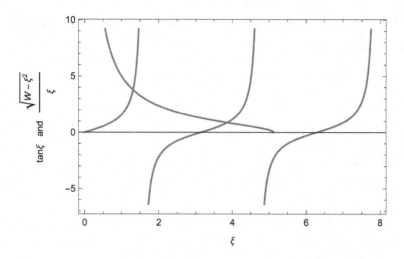

Figure 6.8 Plot of $\tan\xi$ and $\dfrac{\sqrt{W-\xi^2}}{\xi}$. vs. ξ. The solution for the even energies are the points where the curves meet. In this case there are 2 solutions for the even energies, corresponding to the ground and 2nd excited states.

```
In[26]:= ClearAll["Global`*"];

       m = ⌜electron PARTICLE⌝⌜mass⌝ ; V = 100 eV ; L = 0.2 nm ;

           m V L²
       W = ─────── ;
            2 ℏ²

       $Assumptions = {x > 0};

                         √W - x²
       s = FindRoot[Tan[x] - ─────── == 0, {x, 1}];
                            x

       ξ = x /. s[[1]];

                                     2 ξ² ℏ²
       NumberForm[UnitConvert[ ───────, eV ], 3]
                                      m L²

                         √W - x²
       s = FindRoot[Tan[x] - ─────── == 0, {x, 2}];
                            x

       ξ = x /. s[[1]];

                                     2 ξ² ℏ²
       NumberForm[UnitConvert[ ───────, eV ], 3]
                                      m L²

Out[30]//NumberForm=
       6.56 eV

Out[33]//NumberForm=
       56.7 eV
```

Figure 6.9 shows that there are two solutions for the energy where $\tan\xi$ and $\frac{\sqrt{W-\xi^2}}{\xi}$ are equal.

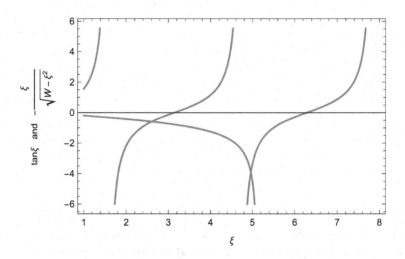

Figure 6.9 Plot of $\tan\xi$ and $-\frac{\xi}{\sqrt{W-\xi^2}}$. vs. ξ. The solution for the odd energies are the points where the curves meet. In this case there are 2 solutions for the odd energies, corresponding to the 1st and 3rd excited states.

Example 6.21 Find the energies of the odd states.

```
In[34]:= s = FindRoot[Tan[x] + ──────── == 0, {x, 2}];
                               √(W - x²)

        ξ = x /. s[[1]]; m = | electron PARTICLE | | mass |;

                         ⎡ 2 ξ² ℏ ²        ⎤
        NumberForm⎢UnitConvert⎢───────, eV⎥, 3⎥
                         ⎣ m L²            ⎦

                                    x
        s = FindRoot⎢Tan[x] + ──────── == 0, {x, 4.9}⎥;
                               √(W - x²)

        ξ = x /. s[[1]];

                         ⎡ 2 ξ² ℏ ²        ⎤
        NumberForm⎢UnitConvert⎢───────, eV⎥, 3⎥
                         ⎣ m L²            ⎦
```

Out[36]//NumberForm=
        25.9 eV

Out[39]//NumberForm=
        93.8 eV

6.9.4   Wave Functions

The wave functions are plotted in fig. 6.10 and the squares are plotted in fig. 6.11.

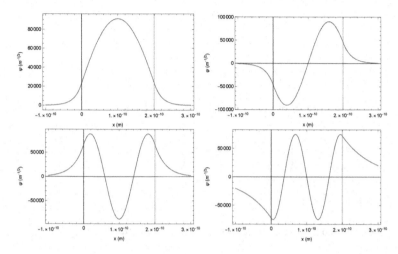

Figure 6.10   Wave functions for the 4 bound states.

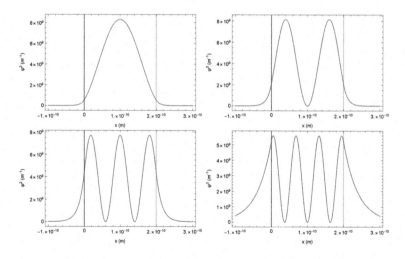

Figure 6.11   Wave functions squared for the 4 bound states.

The wave function squared is normalized to unity.

Example 6.22 Calculate the normalization constant for the ground state and

evaluate the probability for the electron to be outside the box. The ground
state energy has been saved as E_g (see Ex. 6.20).

In[40]:= $E = 6.56$ eV ;

$$\alpha = \text{UnitConvert}\left[\frac{\sqrt{2\,m\,(V - E)}}{\hbar}\right];$$

$$k = \text{UnitConvert}\left[\frac{\sqrt{2\,m\,E}}{\hbar}\right];$$

$$\text{area} = 2\,\text{Integrate}\left[e^{2\,\alpha\,m\,x},\,\{x,\,-\infty,\,0\}\right] +$$

$$\text{Integrate}\left[\left(\frac{1}{\cos\left[\frac{kL}{2}\right]}\,\cos\left[k\,m\,\left(x - \frac{L}{2\,m}\right)\right]\right)^2,\,\left\{x,\,0,\,\frac{L}{m}\right\}\right];$$

$$A = \sqrt{\frac{1}{\text{area}}};$$

$$\text{NumberForm}\left[2\,\text{Integrate}\left[\left(A\,e^{-\alpha\,x}\right)^2,\,\{x,\,0,\,\infty\}\right],\,3\right]$$

Out[44]//NumberForm=

    0.011

Since the wave function extends beyond the potential boundary with exponential decay, the ground state energy may be approximated as that due to an infinite potential of a box of size $L + \frac{2}{\alpha}$.

Example 6.23 Estimate the ground state energy from the infinite potential result,

In[45]:= $\mathtt{NumberForm}\left[\mathtt{UnitConvert}\left[\dfrac{\pi^2\,\hbar^2}{2\left(L + \frac{2}{\alpha}\right)^2 m},\ \mathtt{eV}\right],\ 2\right]$

Out[45]//NumberForm=
       6.5 eV

The estimation is very good as long as $E \ll V$.

CHAPTER 7

Quantum Harmonic Oscillator

The harmonic oscillator is defined by a quadratic potential,

$$U = \frac{1}{2}m\omega^2 x^2,$$

where ω is the angular frequency of an oscillating mass m.

The Schrödinger equation is

$$\frac{d^2\psi}{dx^2} = -\frac{2mE}{\hbar^2}\psi + \frac{1}{2}m\omega^2 x^2\psi.$$

Many important problems in physics reduce to that of a quantum harmonic oscillator because the leading term of a potential near its minimum is quadratic.

7.1 GROUND STATE

7.1.1 Wave Function

The ground state wave function is of the form

$$\psi = Ae^{-\alpha x^2}.$$

where A and α are constants. This is a Gaussian distribution with the parameter α giving the width of the wave function. It is the simplest function whose second derivative gives the same function back again with a minus sign plus

DOI: 10.1201/9781003395515-7

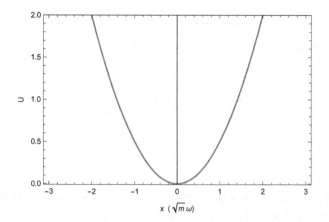

x ($\sqrt{m}\,\omega$)

Figure 7.1 The potential for the harmonic oscillator is quadratic.

a term which has x^2 times the function. The Schrödinger equation may be written

$$\frac{1}{\psi}\frac{d^2\psi}{dx^2} = -\frac{2mE}{\hbar^2} + \frac{1}{2}m\omega^2 x^2.$$

Example 7.1 Calculate $\frac{1}{\psi}\frac{d^2\psi}{dx^2}$.

In[1]:= ψ = A e$^{-\alpha x^2}$; $\dfrac{1}{\psi}$ $\partial_x \partial_x \psi$ // Simplify

Out[1]= $2\,\alpha\,\left(-1 + 2\,x^2\,\alpha\right)$

Comparing separately both x^2 and x terms gives α and E.

Example 7.2 Solve for E and α.

In[2]:= \$Assumptions = {α > 0, m > 0, ω > 0, $\hbar$ > 0};

Solve$\left[-2\,\alpha = -\dfrac{(2\,m\,E)}{\hbar^2} \,\&\&\, 4\,\alpha^2 = \dfrac{m^2\,\omega^2}{\hbar^2},\ \{E,\,\alpha\},\ \text{Reals}\right]$ // Simplify

Out[2]= $\left\{\left\{E \to \dfrac{\omega\,\hbar}{2},\ \alpha \to \dfrac{m\,\omega}{2\,\hbar}\right\}\right\}$

7.1.2 Energy

Knowing the value of α, one can directly substitute ψ into the Schrödinger equation, verifying the solution.

Example 7.3 Solve the Schrödinger equation to get the ground state energy.

In[3]:= `$Assumptions = {m > 0, ω > 0, ℏ > 0};`

$$\psi = A\,e^{-\alpha x^2}; \quad \alpha = \frac{m\,\omega}{2\,\hbar};$$

$$\text{Solve}\left[\partial_x \partial_x \psi == -\frac{2\,m\,E}{\hbar^2}\,\psi + \frac{m^2\,\omega^2\,x^2}{\hbar^2}\,\psi,\ \{E\},\ \mathbb{R}\right]$$

Out[4]= $\left\{\left\{E \to \dfrac{\omega\,\hbar}{2}\right\}\right\}$

It is a general and noteworthy feature of quantum systems that the lowest energy is not zero. Compare to a particle in a box (6.3).

7.1.3 Normalization

The constant A is determined from the normalization condition,

$$\int_0^\infty \psi^2 dx = 1.$$

Example 7.4 Find the normalization constant A.

In[5]:= $A = \dfrac{1}{\sqrt{\int_{-\infty}^{\infty} e^{-2\alpha x^2}\, dx}}$

Out[5]= $\dfrac{1}{\pi^{1/4}\sqrt{\dfrac{\hbar}{\sqrt{m\,\omega\,\hbar}}}}$

7.1.4 Quantum Tunneling

A classical oscillator has maximum position ($x_{\max}$) given by conservation of energy,

$$\frac{1}{2}m\omega^2 x_{\max}^2 = E.$$

For the ground state energy ($\hbar\omega/2$) this would be

$$x_{\max} = \sqrt{\frac{\hbar}{m\omega}}$$

(see fig. 7.2).

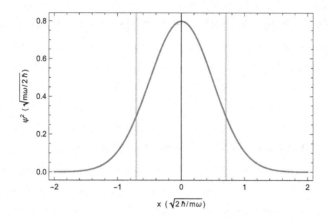

Figure 7.2 Wave function squared for the ground state with the classical limits indicated by vertical lines (left and right). The square of a Gaussian is also a Gaussian.

Example 7.5 Calculate the probability that the quantum oscillator is found beyond the classical limit (on either side).

In[6]:= $N\left[2 \int_{\sqrt{\frac{\hbar}{m\omega}}}^{\infty} \left(A\, e^{-\alpha x^2}\right)^2 dx\right]$

Out[6]= 0.157299

Examination of a plot of ψ^2 (fig. 7.2) shows that the probability peaks at $x = 0$. This is opposite to that of a classical oscillator that spends the least time at $x = 0$ where it is moving the fastest.

Example 7.6 Calculate the probability that the quantum oscillator is found within the center half of the classical range.

In[7]:= $N\left[\int_{-\frac{1}{2}\sqrt{\frac{\hbar}{m\omega}}}^{\frac{1}{2}\sqrt{\frac{\hbar}{m\omega}}} \left(A\, e^{-\alpha x^2}\right)^2 dx\right]$

Out[7]= 0.5205

In 7.4.2 the classical probability is calculated to be $\frac{1}{3}$.

7.1.5 Uncertainty Principle

Agreement with the uncertainty principle may be checked by calculating the expectation values of x^2 and p^2 and then taking the square root of the product.

Example 7.7 Calculate the expectation value of x^2.

In[8]:= $\int_{-\infty}^{\infty} \psi^2 \, x^2 \, dx$

Out[8]= $\dfrac{\hbar}{2\,m\,\omega}$

Example 7.8 Calculate the expectation value of p^2.

In[9]:= $\int_{-\infty}^{\infty} \psi \, (-i\,\hbar\,\partial_x \, (-i\,\hbar\,\partial_x\psi)) \, dx$

Out[9]= $\dfrac{m\,\omega\,\hbar}{2}$

Example 7.9 Calculate the product $\Delta x \Delta p$.

In[10]:= $\sqrt{\dfrac{\hbar}{2\,m\,\omega}} \ \sqrt{\dfrac{m\,\omega\,\hbar}{2}}$ // Simplify

Out[10]= $\dfrac{\hbar}{2}$

The absolute limit of the uncertainty $\Delta x \Delta p$ has been achieved because the wave function squared is a Gaussian distribution.

7.2 FIRST EXCITED STATE

7.2.1 Wave Function

The first excited state wave function is of the form

$$\psi = Axe^{-\alpha x^2}.$$

The normalization constant A will differ from that of the ground state. This function is a solution because it's second derivative again gives the same function back again with a minus sign plus a term which has x^2 times the function.

Example 7.10 Calculate $\frac{1}{\psi}\frac{d^2\psi}{dx^2}$.

In[11]:= $\psi = A \, x \, e^{-\alpha x^2}; \ \dfrac{\partial_x \partial_x \psi}{\psi}$ // Simplify

Out[11]= $\dfrac{m\,\omega\,\left(m\,x^2\,\omega - 3\,\hbar\right)}{\hbar^2}$

The x^2 term of $\frac{1}{\psi} \frac{d^2 \psi}{dx^2}$ is seen to be identical to that of the ground state, therefore, the solution for α is the same. The constant term is 3 times that in the ground state, giving 3 times the energy.

7.2.2 Energy

Example 7.11 Solve the Schrödinger equation to get the energy of the first excited state.

In[12]:= $\psi = A \, x \, e^{-\alpha x^2}$; Solve$\left[\partial_x \partial_x \psi == -\frac{2 \, m \, E}{\hbar^2} \psi + \frac{m^2 \, \omega^2 \, x^2}{\hbar^2} \psi, \, \{E\}, \, \mathbb{R}\right]$

Out[12]=

$$\left\{\left\{E \to \frac{3 \, \omega \, \hbar}{2}\right\}\right\}$$

7.2.3 Normalization

Example 7.12 Find the normalization constant A.

In[13]:= $A = \dfrac{1}{\sqrt{\int_{-\infty}^{\infty} x^2 \, e^{-2\alpha x^2} \, dx}}$

Out[13]= $\dfrac{\sqrt{2}}{\pi^{1/4} \sqrt{\left(\frac{\hbar}{m \, \omega}\right)^{3/2}}}$

7.2.4 Quantum Tunneling

The classical limit of $x_{\max}$ for the first excited state ($E = 3\hbar\omega/2$) is

$$x_{\max} = \sqrt{\frac{3\hbar}{m\omega}}.$$

Example 7.13 Calculate the probability that the quantum oscillator is found beyond the classical limit.

In[14]:= $N\left[2 \int_{\sqrt{\frac{3\hbar}{m\omega}}}^{\infty} \left(A \, x \, e^{-\alpha x^2}\right)^2 \, dx\right]$

Out[14]=

0.11161

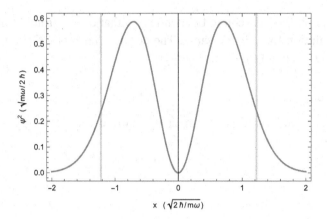

Figure 7.3 The wave function squared for the 1st excited state is shown with the classical limits indicated by vertical lines (left and right).

7.2.5 Uncertainty Principle

Example 7.14 Calculate the expectation value of x^2.

In[15]:= $\int_{-\infty}^{\infty} \psi^2 \, x^2 \, dx$

Out[15]=

$$\frac{3\,\hbar^2}{4\,m^2\,\omega^2}$$

Example 7.15 Calculate the expectation value of p^2.

In[16]:= $\int_{-\infty}^{\infty} \psi \, (-i\,\hbar\,\partial_x \, (-i\,\hbar\,\partial_x\psi)) \, dx$

Out[16]=

$$\frac{3\,\hbar^2}{4}$$

Example 7.16 Calculate the product $\Delta x \Delta p$.

In[17]:= $\sqrt{\dfrac{3\,\hbar}{2\,m\,\omega}} \, \sqrt{\dfrac{3\,m\,\omega\,\hbar}{2}}$ // Simplify

Out[17]=

$$\frac{3\,\hbar}{2}$$

7.3   GENERAL SOLUTION

The normalized wave functions of the quantum harmonic oscillator may be generally expressed as

$$\psi_n = \frac{1}{\sqrt{2^n n!}}(\frac{2\alpha}{\pi})^{1/4} H_n(\sqrt{2\alpha}x)e^{-ax^2},$$

where $H_n(\sqrt{2\alpha}x)$ is a Hermite polynomial.

7.3.1   Hermite Polynomials

Example 7.17 Get the Hermite polynomial and calculate $\frac{1}{\psi}\frac{d^2\psi}{dx^2}$ for $n = 0$.

In[18]:= ψ = A HermiteH$\left[n, \sqrt{2\alpha} \ x\right] e^{-\alpha x^2}$ /. n → 0

$\dfrac{\partial_x \partial_x \psi}{\psi}$ // Simplify

Out[18]=

$$\frac{\sqrt{2} \ e^{-\frac{m x^2 \omega}{2\hbar}}}{\pi^{1/4} \ \sqrt{\left(\frac{\hbar}{m\omega}\right)^{3/2}}}$$

Out[19]=

$$\frac{m\omega \left(m x^2 \omega - \hbar\right)}{\hbar^2}$$

Example 7.18 Get the Hermite polynomial and calculate $\frac{1}{\psi}\frac{d^2\psi}{dx^2}$ for $n = 1$.

In[20]:= ψ = A HermiteH$\left[n, \sqrt{2\alpha} \ x\right] e^{-\alpha x^2}$ /. n → 1

$\dfrac{\partial_x \partial_x \psi}{\psi}$ // Simplify

Out[20]=

$$\frac{2 \sqrt{2} \ e^{-\frac{m x^2 \omega}{2\hbar}} \ x \ \sqrt{\frac{m\omega}{\hbar}}}{\pi^{1/4} \ \sqrt{\left(\frac{\hbar}{m\omega}\right)^{3/2}}}$$

Out[21]=

$$\frac{m\omega \left(m x^2 \omega - 3\hbar\right)}{\hbar^2}$$

Example 7.19 Get the Hermite polynomial and calculate $\frac{1}{\psi}\frac{d^2\psi}{dx^2}$ for $n = 2$.

In[22]:= ψ = A HermiteH$\left[n, \sqrt{2\,\alpha}\ x\right]$ e$^{-\alpha x^2}$ /. n → 2

$\dfrac{\partial_x\partial_x\psi}{\psi}$ // Simplify

Out[22]=

$$\frac{\sqrt{2}\ e^{-\frac{m x^2 \omega}{2\hbar}}\left(-2 + \frac{4\,m\,x^2\,\omega}{\hbar}\right)}{\pi^{1/4}\ \sqrt{\left(\frac{\hbar}{m\,\omega}\right)^{3/2}}}$$

Out[23]=

$$\frac{m\,\omega\,\left(m\,x^2\,\omega - 5\,\hbar\right)}{\hbar^2}$$

Example 7.20 Get the Hermite polynomial and calculate $\frac{1}{\psi}\frac{d^2\psi}{dx^2}$ for $n = 3$.

In[24]:= ψ = A HermiteH$\left[n, \sqrt{2\,\alpha}\ x\right]$ e$^{-\alpha x^2}$ /. n → 3

$\dfrac{\partial_x\partial_x\psi}{\psi}$ // Simplify

Out[24]=

$$\frac{\sqrt{2}\ e^{-\frac{m x^2 \omega}{2\hbar}}\left(-12\,x\,\sqrt{\frac{m\,\omega}{\hbar}} + 8\,x^3\,\left(\frac{m\,\omega}{\hbar}\right)^{3/2}\right)}{\pi^{1/4}\ \sqrt{\left(\frac{\hbar}{m\,\omega}\right)^{3/2}}}$$

Out[25]=

$$\frac{m\,\omega\,\left(m\,x^2\,\omega - 7\,\hbar\right)}{\hbar^2}$$

Example 7.21 Get the Hermite polynomial and calculate $\frac{1}{\psi}\frac{d^2\psi}{dx^2}$ for $n = 4$.

In[26]:= ψ = A HermiteH$\left[n, \sqrt{2\,\alpha}\ x\right]$ e$^{-\alpha x^2}$ /. n → 4

$\dfrac{\partial_x\partial_x\psi}{\psi}$ // Simplify

Out[26]=

$$\frac{\sqrt{2}\ e^{-\frac{m x^2 \omega}{2\hbar}}\left(12 + \frac{16\,m^2\,x^4\,\omega^2}{\hbar^2} - \frac{48\,m\,x^2\,\omega}{\hbar}\right)}{\pi^{1/4}\ \sqrt{\left(\frac{\hbar}{m\,\omega}\right)^{3/2}}}$$

Out[27]=

$$\frac{m\,\omega\,\left(m\,x^2\,\omega - 9\,\hbar\right)}{\hbar^2}$$

7.3.2 Normalization

Example 7.22 Check the normalization of the first 20 states.

In[28]:= $\alpha = \dfrac{m\,\omega}{2\,\hbar}$; $\psi = \dfrac{\left(\frac{2\,\alpha}{\pi}\right)^{1/4}}{\sqrt{2^n\,n!}}\; e^{-\alpha x^2}\; \text{HermiteH}\left[n,\; \sqrt{2\,\alpha}\;x\right]$;

$\text{Table}\left[\displaystyle\int_{-\infty}^{\infty}\psi^2\,dx,\; \{n,\,0,\,19\}\right]$

Out[29]=

$\{1,\,1,\,1,\,1,\,1,\,1,\,1,\,1,\,1,\,1,\,1,\,1,\,1,\,1,\,1,\,1,\,1,\,1,\,1,\,1\}$

7.3.3 Energy

The oscillator energy is given by

$$E = \left(n + \frac{1}{2}\right)\hbar\omega$$

for $n = 0, 1, 2,$ *etc.*

Example 7.23 Calculate the energies of the first 20 states.

In[30]:= $\$\text{Assumptions} = \{m > 0,\; \omega > 0,\; \hbar > 0\}$; $\alpha = \dfrac{m\,\omega}{2\,\hbar}$;

$\text{MatrixForm}\Big[\text{Table}\Big[\text{Table}\Big[\psi = A\,\text{HermiteH}\left[n,\; \sqrt{2\,\alpha}\;x\right]\,e^{-\alpha x^2}$;

$\text{Solve}\left[\partial_x\,\partial_x\,\psi = -\dfrac{2\,m\,E}{\hbar^2}\,\psi + \dfrac{m^2\,\omega^2\,x^2}{\hbar^2}\,\psi,\; \{E\},\; \mathbb{R}\right]$,

$\{n,\, 0 + 4\,j,\, 3 + 4\,j\}\Big],\; \{j,\, 0,\, 4\}\Big]\Big]$

Out[31]//MatrixForm=

$$\begin{pmatrix}
\left(E \to \frac{\omega\hbar}{2}\right) & \left(E \to \frac{3\,\omega\hbar}{2}\right) & \left(E \to \frac{5\,\omega\hbar}{2}\right) & \left(E \to \frac{7\,\omega\hbar}{2}\right) \\[4pt]
\left(E \to \frac{9\,\omega\hbar}{2}\right) & \left(E \to \frac{11\,\omega\hbar}{2}\right) & \left(E \to \frac{13\,\omega\hbar}{2}\right) & \left(E \to \frac{15\,\omega\hbar}{2}\right) \\[4pt]
\left(E \to \frac{17\,\omega\hbar}{2}\right) & \left(E \to \frac{19\,\omega\hbar}{2}\right) & \left(E \to \frac{21\,\omega\hbar}{2}\right) & \left(E \to \frac{23\,\omega\hbar}{2}\right) \\[4pt]
\left(E \to \frac{25\,\omega\hbar}{2}\right) & \left(E \to \frac{27\,\omega\hbar}{2}\right) & \left(E \to \frac{29\,\omega\hbar}{2}\right) & \left(E \to \frac{31\,\omega\hbar}{2}\right) \\[4pt]
\left(E \to \frac{33\,\omega\hbar}{2}\right) & \left(E \to \frac{35\,\omega\hbar}{2}\right) & \left(E \to \frac{37\,\omega\hbar}{2}\right) & \left(E \to \frac{39\,\omega\hbar}{2}\right)
\end{pmatrix}$$

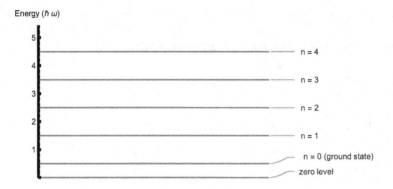

Figure 7.4 The energy levels of the quantum harmonic oscillator are equally spaced.

7.3.4 Wave Functions

The ground state wave function has a single "bump" and each excited state has one bump more than the state below it.

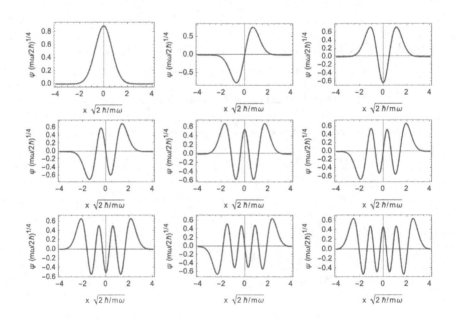

Figure 7.5 The wave functions for the first 9 states are shown.

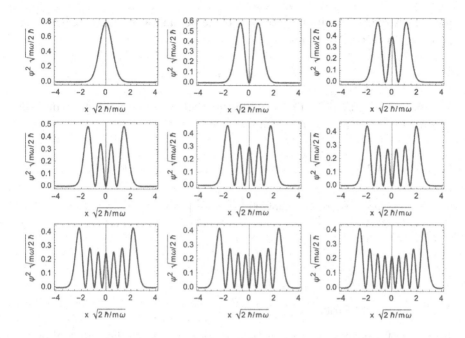

Figure 7.6   The wave functions squared for the first 9 states are shown.

7.4   CORRESPONDENCE PRINCIPLE

7.4.1   Large Quantum Numbers

For large quantum numbers, the solution must approach that of the classical harmonic oscillator. The (angular) frequency of a simple pendulum consisting of a mass m on a string of length L is

$$\omega = \sqrt{\frac{g}{L}},$$

where g is the acceleration of gravity. If the pendulum swings to a height H above its lowest point, its energy is

$$E = mgH.$$

Example 7.24 Calculate the quantum number of a simple pendulum with $m = 0.1$ kg, $L = 1$ m, and $H = 1$ mm.

In[32]:= `m = 0.1 kg ; L = 1 m ; H = 1 mm ; ω =` $\sqrt{\dfrac{g}{L}}$ `; E = m g H;`

`n = NumberForm`$\left[\text{UnitConvert}\left[\dfrac{E}{\hbar\,\omega} - \dfrac{1}{2}\right], 1\right]$

Out[33]//NumberForm=

$3. \times 10^{30}$

7.4.2 Classical Probability

In one-half period $(T/2)$, a classical particle traverses the entire allowed range. The classical probability distribution (P) is

$$P = \frac{1}{\frac{T}{2}v(x)},$$

where the speed is given by

$$v(x) = \sqrt{\frac{2K}{m}} = \sqrt{\frac{2}{m}(E - U)} = \sqrt{\frac{2}{m}\left(\frac{1}{2}m\omega^2 x_{\text{max}}^2 - \frac{1}{2}m\omega^2 x^2\right)}$$

Using $T = 2\pi/\omega$,

$$P = \frac{1}{\pi\sqrt{x_{\text{max}}^2 - x^2}}.$$

The classical maximum displacement corresponding to energy with quantum number n is

$$x_{\text{max}} = \sqrt{\frac{(2n + 1)m\omega}{\hbar}}.$$

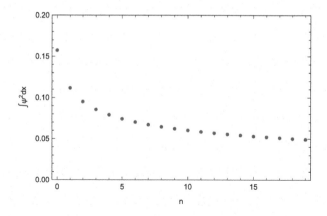

Figure 7.7 The probability for a quantum oscillator to be found beyond the classical limit is shown *vs. n*.

Example 7.25 Calculate the classical probability that the particle is in the center half of the oscillation range.

In[34]:= **ClearAll["Global`*"];**

$Assumptions = {n > 0, m > 0, ℏ > 0, ω > 0};

$$\text{Integrate}\left[\frac{1}{\pi \sqrt{\frac{(2n+1)\,m\,\omega}{2\,\hbar} - x^2}} , \right.$$

$$\left. \left\{x, -\frac{1}{2}\sqrt{\frac{(2n+1)\,m\,\omega}{2\,\hbar}}, \frac{1}{2}\sqrt{\frac{(2n+1)\,m\,\omega}{2\,\hbar}}\right\}\right]$$

Out[35]= $\dfrac{1}{3}$

Example 7.26 Calculate the probability that a quantum oscillator is in the center half of the classical oscillation range for *n* = 50.

In[36]:= **n = 50;**

$$\psi = 1/\text{Sqrt}\left[2^n\,n!\right]\left(\frac{2}{\pi}\right)^{1/4} e^{-x^2}\,\text{HermiteH}\left[n, \sqrt{2}\,x\right];$$

$$\text{NIntegrate}\left[\psi^2, \left\{x, -\frac{1}{2}\frac{\sqrt{2n+1}}{\sqrt{2}}, \frac{1}{2}\frac{\sqrt{2n+1}}{\sqrt{2}}\right\}\right]$$

Out[36]=

0.336215

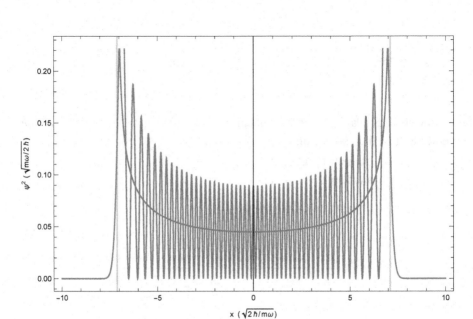

Figure 7.8 The wave function squared for $n=50$ is shown together with the classical probability distribution.

CHAPTER 8

Hydrogen Atom

The Bohr model (5) is highly successful in explaining many features of the hydrogen atom. The next level of sophistication is to solve the Schrödinger equation,

$$-\frac{\hbar^2}{2m}\nabla^2\psi - \frac{e^2}{4\pi\varepsilon_0 r}\psi = E\psi.$$

The solution may be obtained by the technique of separation of variables,

$$\psi(r,\theta\phi) = R(r)Y(\theta,\phi),$$

resulting in radial

$$\frac{1}{r^2}\frac{d}{dr}(r^2\frac{dR}{dr}) + \frac{2m}{\hbar^2}(E - V - \frac{\hbar^2\ell(\ell+1)}{2mr^2})R = 0$$

and angular

$$\frac{1}{\sin\theta}\frac{d}{d\theta}(\sin\theta\frac{dY}{d\theta} + [\ell(\ell+1) - \frac{m_\ell^2}{\sin^2\theta}]Y = 0$$

equations. The integers ℓ and m_ℓ appear as part of the separation process. The radial solutions are of the form

$$R = f(r)e^{-\frac{r}{na}},$$

where $f(r)$ is a polynomial and n is a positive integer. The angular solution is the spherical harmonic. The quantity

$$L = \hbar^2\ell(\ell+1)$$

has an interpretation as orbital angular momentum with

$$L_z = m_\ell\hbar$$

DOI: 10.1201/9781003395515-8

being its "z component", the possible value of **L** along any given direction. The integer n appears in the solution for allowed energies from the second derivative in the radial equation.

$$E = \frac{E_0}{n^2},$$

where E_0 is the ground state energy.

8.1 GROUND STATE

The ground state wave function does not have any angular dependence and may be written

$$\psi = Ae^{-r/a}.$$

It is the simplest function whose Laplacian gives the terms with function reproduced and $1/r$ times the function. Its quantum numbers are $(n, \ell, m_\ell) = (1,0,0)$. In spectroscopic notation, this is referred to as the $1s$ state.

8.1.1 Solution

The Schrödinger equation is

$$\frac{1}{\psi}\nabla^2\psi = -\frac{2me^2}{4\pi\varepsilon_0\hbar^2 r} - \frac{2mE}{\hbar^2}.$$

Example 8.1 Calculate $\frac{1}{\psi}\nabla^2\psi$.

In[1]:= ψ = A e$^{-r/a}$; $\frac{1}{\psi}$ Laplacian[ψ, {r, θ, ϕ}, "Spherical"] // Simplify

Out[1]= $\dfrac{-2a+r}{a^2 r}$

Comparing, separately, the terms containing r and the constant terms gives a and E.

Example 8.2 Solve for E and a.

In[2]:= Solve$\left[\dfrac{1}{a^2} == -\dfrac{(2\,m\,E)}{\hbar^2}\ \&\&\ -\dfrac{2}{a} == -\dfrac{2\,m\,e^2}{4\,\pi\,\varepsilon_0\,\hbar^2},\ \{E,\,a\},\ \text{Reals}\right]$ // Simplify

Out[2]= $\left\{\left\{E \to -\dfrac{e^4\,m}{32\,\pi^2\,\hbar^2\,\varepsilon_0^2},\ a \to \dfrac{4\,\pi\,\hbar^2\,\varepsilon_0}{e^2\,m}\right\}\right\}$

The parameter a appearing in the exponential of the wave function is seen to be the Bohr radius as calculated in 5.1.

8.1.2 Normalized Wave Functions

The normalized hydrogen atom wave functions are directly accessible in Mathematica (see A.7) and can be readily verified as solutions of the Schrödinger equation.

Example 8.3 Get the hydrogen atom ground state wave function in terms of the Bohr radius a.

In[3]:= ψ = [■] HydrogenWavefunction ✦ [{1, 0, 0}, a, {r, θ, φ}]

Out[3]= $\dfrac{\sqrt{\frac{1}{a^3}}\ e^{-\frac{r}{a}}}{\sqrt{\pi}}$

Example 8.4 Show that this function satisfies the Schrödinger equation and calculate the energy.

In[4]:= $Assumptions = {a > 0, m > 0, ε₀ > 0};

Solve$\Big[$

$\quad -\dfrac{\hbar^2}{2\,m}$ Laplacian[ψ, {r, θ, φ}, "Spherical"] ==

$\quad \Big(E + \dfrac{e^2}{4\,\pi\,\varepsilon_0\,r}\Big)\,\psi\ /.\ a \to \dfrac{4\,\pi\,\hbar^2\,\varepsilon_0}{e^2\,m},\ E,\ Reals\Big]$ // Simplify

Out[5]= $\Big\{\Big\{E \to -\dfrac{e^4\,m}{32\,\pi^2\,\hbar^2\,\varepsilon_0^2}\Big\}\Big\}$

Example 8.5 Evaluate the ground state energy numerically.

In[6]:= m = [electron PARTICLE] [mass];

$\quad N\Big[UnitConvert\Big[-\dfrac{e^4\,m}{32\,\pi^2\,\hbar^2\,\varepsilon_0^2},\ eV\Big],\ 3\Big]$

Out[7]= -13.6 eV

Example 8.6 Check the normalization by direct integration.

$$\text{In[8]:=} \int_0^\infty \left(\int_0^{2\pi} \left(\int_0^\pi r^2 \, \text{Sin[}\theta\text{] Conjugate[}\psi\text{] }\psi \, d\theta \right) d\phi \right) dr$$

Out[8]= 1

8.1.3 Radial Probability

The probability of finding the electron per volume at a given location is given by the square of the wave function,

$$\frac{dP}{dV} = |\psi|^2 .$$

The integration over the angular variables gives 4π since there is no angular dependence. This leads to a radial probability distribution,

$$\frac{dP}{dr} = 4\pi r^2 |\psi|^2 = \frac{4r^2}{a^3} e^{-2r/a}.$$

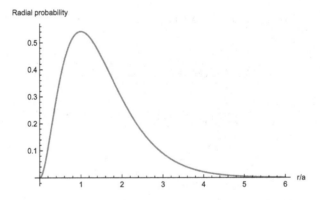

Figure 8.1 Radial probability for the ground state of hydrogen is plotted in units of the Bohr radius.

The most-probable radius is found by setting the derivative of the probability function equal to zero and then solving for r. It turns out to be the Bohr radius as can be seen from Fig. 8.1.

Example 8.7 Calculate the most probable distance for the electron to be from the proton.

In[9]:= `$Assumptions = {r > 0, a > 0}; Solve[∂_r (4 π r² ψ²) == 0, r, ℝ]`

Out[9]= `{{r → a}}`

Example 8.8 Calculate the average distance.

In[10]:= `$Assumptions = a > 0; ∫₀^∞ 4 π r³ ψ² dr`

Out[10]=
$$\frac{3\,a}{2}$$

Example 8.9 Calculate the root-mean-square distance.

In[11]:= `√(∫₀^∞ 4 π r⁴ ψ² dr) // Simplify`

Out[11]=
$$\sqrt{3}\,a$$

It is seen that the most probable (r_{mp}), average (r_{ave}), and rms (r_{rms}) distances are related by

$$r_{\mathrm{mp}} < r_{\mathrm{ave}} < r_{\mathrm{rms}}.$$

Figure 8.1 shows that there is a substantial probability of finding the electron beyond the Bohr radius.

Example 8.10 Calculate probability that the electron is found beyond the Bohr radius.

In[12]:= `∫_a^∞ 4 π r² ψ² dr`

Out[12]=
$$\frac{5}{e^2}$$

This is about 68%.

The electron has only a very tiny probability of being found inside the proton, whose size is about 1 fm. At such a small distance, the exponential part of ψ^2 is essentially unity, and it can either be expanded (or just set to unity and pulled outside the integral).

Example 8.11 Estimate the probability that the electron may be found inside the proton.

In[13]:= **d = UnitConvert$\left[\dfrac{1 \text{ fm}}{a_\theta}\right]$;**

ScientificForm$\left[\displaystyle\int_\theta^{d\,a} 4\,\pi\,r^2\,\left(\dfrac{1 - \frac{2\,r}{a}}{a^3\,\pi}\right)\,dr,\ 1\right]$

Out[13]//ScientificForm=
 $9. \times 10^{-15}$

8.1.4 Consistency with the Uncertainty Principle

The expectation value for the kinetic energy can be evaluated with use of the momentum operator,

$$\langle K \rangle = \int_0^\infty 4\pi r^2 \psi^* \left(-\frac{\hbar^2}{2m}\right) \nabla^2 \psi dr.$$

Example 8.12 Calculate the average kinetic energy and give the numerical value.

In[15]:= **Clear[m];**

K = $\displaystyle\int_\theta^\infty 4\,\pi\,r^2\,\psi\,\left(-\dfrac{\hbar^2}{2\,m}\ \text{Laplacian}[\,\psi,\ \{r,\ \theta,\ \phi\},\ \text{"Spherical"}]\right)dr$

N$\left[\text{UnitConvert}[K,\ eV]\ /.\right.$
 $\left.\left\{\hbar \rightarrow \hbar,\ m \rightarrow \boxed{\text{electron}\ \text{PARTICLE}}\,\boxed{\text{mass}}\right],\ a \rightarrow a_\theta\right\},\ 3\right]$

Out[15]= $\dfrac{\hbar^2}{2\,a^2\,m}$

Out[16]= $13.6\ eV$

To check the uncertainty principle, one may assign a value

$$\Delta p = \sqrt{2mK}$$

due to lack of knowledge of the direction or sign of p, as we only know its average square. The rms radial position may be used to give

$$\Delta x = \sqrt{3}a.$$

Example 8.13 Check the uncertainty principle for the ground state of hydrogen.

```
In[17]:= $Assumptions = {ℏ > 0, a > 0};
        Δx = √3 a;
        Δp = √(2 m K);
        Δx Δp // Simplify
```

```
Out[17]= √3 ℏ
```

8.2 FIRST EXCITED STATES

The first excited state has $n = 2$ and 2 possibilities for the quantum number ℓ, 0 or 1, referred to as the $2s$ and $2p$ states, respectively. The $\ell = 1$ state has three possibilities for m_ℓ, $-1, 0,$ or 1. Thus, the Schrödinger equation gives 4 solutions.

8.2.1 Wavefunctions

Example 8.14 Get the n=2 wave functions.

```
In[18]:= s = ResourceFunction["HydrogenWavefunction"][{2, 0, 0},
            a, {r, θ, ϕ}]
        p = Table[ResourceFunction[
            "HydrogenWavefunction"][{2, 1, m}, a, {r, θ, ϕ}],
            {m, -1, 1}]
```

$$Out[18]= \frac{\sqrt{\frac{1}{a^3}}\, e^{-\frac{r}{2a}} \left(2 - \frac{r}{a}\right)}{4\sqrt{2\pi}}$$

$$Out[19]= \left\{ \frac{\sqrt{\frac{1}{a^3}}\, e^{-\frac{r}{2a} - i\phi}\, r\, \mathrm{Sin}[\theta]}{8a\sqrt{\pi}} \right. ,$$
$$\left. \frac{\sqrt{\frac{1}{a^3}}\, e^{-\frac{r}{2a}}\, r\, \mathrm{Cos}[\theta]}{4a\sqrt{2\pi}} , -\frac{\sqrt{\frac{1}{a^3}}\, e^{-\frac{r}{2a} + i\phi}\, r\, \mathrm{Sin}[\theta]}{8a\sqrt{\pi}} \right\}$$

8.2.2 Energies

Example 8.15 Show that the $2s$ wave function satisfies the Schrödinger equation and calculate the energy.

In[20]:= **$Assumptions = {a > 0, m > 0, ε₀ > 0}; ψ = s;**

$$\text{Solve}\left[-\frac{\hbar^2}{2\,m}\,\text{Laplacian}[\,\psi,\,\{r,\,\theta,\,\phi\},\,\text{"Spherical"}]\,=\!=\right.$$

$$\left.\left(\text{E} + \frac{e^2}{4\,\pi\,\varepsilon_0\,r}\right)\psi\,/.\,a \to \frac{4\,\pi\,\hbar^2\,\varepsilon_0}{e^2\,m}\,,\,\text{E},\,\text{Reals}\right]\,//\,\text{Simplify}$$

Out[21]= $\left\{\left\{\text{E} \to -\dfrac{e^4\,m}{128\,\pi^2\,\hbar^2\,\varepsilon_0^2}\right\}\right\}$

Example 8.16 Show that the $2p$ wave functions satisfy the Schrödinger equation and calculate the energies.

In[22]:= **Table** $\left[\text{Solve}\left[\psi = p[\![i]\!];\right.\right.$

$$-\frac{\hbar^2}{2\,m}\,\text{Laplacian}[\,\psi,\,\{r,\,\theta,\,\phi\},\,\text{"Spherical"}]\,=\!=$$

$$\left(\text{E} + \frac{e^2}{4\,\pi\,\varepsilon_0\,r}\right)\psi\,/.\,a \to \frac{4\,\pi\,\hbar^2\,\varepsilon_0}{e^2\,m}\,,\,\text{E},\,\text{Reals}\right],\,\{i,\,1,\,3\}\right]\,//$$

Simplify

Out[22]= $\left\{\left\{\left\{\text{E} \to -\dfrac{e^4\,m}{128\,\pi^2\,\hbar^2\,\varepsilon_0^2}\right\}\right\},\right.$

$\left\{\left\{\text{E} \to -\dfrac{e^4\,m}{128\,\pi^2\,\hbar^2\,\varepsilon_0^2}\right\}\right\},\,\left.\left\{\left\{\text{E} \to -\dfrac{e^4\,m}{128\,\pi^2\,\hbar^2\,\varepsilon_0^2}\right\}\right\}\right\}$

All 4 states are seen to have the same energy which is the ground state energy divided by $n^2 = 4$.

8.2.3 Radial Probability Distributions

All of the quantities calculated for the $1s$ state may be easily reproduced for the $2s$ and $2p$ states. The radial probability distribution for the $2s$ state is shown in Fig. 8.2.

Example 8.17 Calculate the average electron distance for the $2s$ state.

In[23]:= **$\psi = s$; $Assumptions = a > 0;** $\displaystyle\int_0^\infty 4\,\pi\,r^2\,\text{Conjugate}[\,\psi]\,r\,\psi\,dr$

Out[23]=

$6\,a$

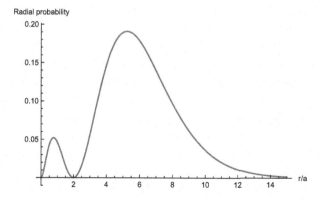

Figure 8.2 Radial probability for the 2s state of hydrogen is plotted in units of the Bohr radius.

The 2p wave functions have angular dependence in both θ and ϕ. Note also that the ϕ parts are represented as complex numbers. Thus, in calculating the radial probability, we must integrate over all angles and also use

$$\left|\psi^2\right| = \psi^*\psi.$$

The general form is

$$\frac{dP}{dr} = \int_0^{2\pi} d\phi \int_0^{\pi} d\theta \, \sin\theta \, r^2 \psi^*\psi.$$

Example 8.18 Compare the radial probabilities for the 2p states.

```
In[24]:=   $Assumptions = {r > 0, a > 0};
           ∫₀²π (∫₀π r² Sin[θ] Conjugate[p[[1]]] p[[1]] dθ) dφ ==
           ∫₀²π (∫₀π r² Sin[θ] Conjugate[p[[2]]] p[[2]] dθ) dφ ==
           ∫₀²π (∫₀π r² Sin[θ] Conjugate[p[[3]]] p[[3]] dθ) dφ

Out[24]=
           True
```

The 2p radial distribution is shown in Fig. 8.3.

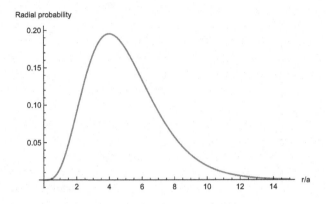

Figure 8.3 Radial probability for the $2p$ state of hydrogen is plotted in units of the Bohr radius.

Example 8.19 Calculate the average distance for the $2p$ states.

In[25]:= $\texttt{Table}\Big[\psi = \texttt{p}[\![\texttt{i}]\!];$

$\texttt{\$Assumptions = a > 0;}$

$\int_0^\pi \left(\int_0^{2\pi} \left(\int_0^\infty \texttt{Sin}[\theta]\ r^2\ \texttt{Conjugate}[\psi]\ r\ \psi\ dr \right) d\phi \right) d\theta,\ \{\texttt{i, 1, 3}\}\Big]$

Out[25]=

$\{5\,a,\ 5\,a,\ 5\,a\}$

8.3 MORE EXCITED STATES

The letters used in the spectroscopic notation are $s, p, d, f, g...$ for $\ell = 0, 1, 2, 3, 4....$ After the letter f, they are just alphabetical order. A schematic of the allowed energy levels looks like

$4s$	$4p$	$4d$	$4f$
$3s$	$3p$	$3d$	
$2s$	$2p$		
$1s$			

with n on the vertical and ℓ horizontal.

Example 8.20 Get the wave function for the 3d state.

In[26]:= ψ = ResourceFunction["HydrogenWavefunction"][{3, 2, 0}, a, {r, θ, ϕ}]

Out[26]= $\dfrac{\sqrt{\dfrac{1}{a^3}}\; e^{-\frac{r}{3a}}\; r^2 \left(-1 + 3\,\text{Cos}[\theta]^2\right)}{81\, a^2\, \sqrt{6\,\pi}}$

Example 8.21 Calculate the average electron distance.

In[27]:= P = 2 π $\displaystyle\int_{0}^{\pi}$ r^2 Sin[θ] ψ^2 dθ;

$\displaystyle\int_{0}^{\infty}$ P r dr

Out[28]=

$\dfrac{21\,a}{2}$

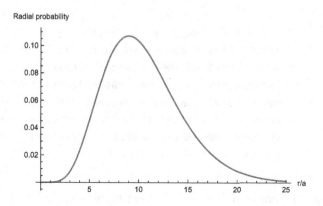

Radial probability

Figure 8.4 The radial probability for the 3d state is shown.

Example 8.22 Calculate probability that the electron is found beyond 9 times the Bohr radius.

In[29]:= $\displaystyle\int_{9a}^{\infty}$ P dr

N[%]

Out[29]=

$\dfrac{1223}{5\,e^6}$

Out[30]=

0.606303

8.4   CORRESPONDENCE PRINCIPLE

At large values of the quantum numbers n and ℓ, one expects a wave function that is sharply peaked at a distance of n^2 times the Bohr radius as the wave function approaches a classical circular orbit.

Example 8.23   Calculate the radial probability for the (100,99,0) state of hydrogen.

In[31]:= ψ = **ResourceFunction**["HydrogenWavefunction"][{100, 99, 0},
        a, {r, θ, φ}];
        2 π **Integrate**[r² **Sin**[θ] **Conjugate**[ψ] ψ, {θ, 0, π}]

Out[31]= $\left(e^{-\frac{r+\mathrm{Conjugate}[r]}{100\,a}}\ r^{101}\ \mathrm{Conjugate}[r]^{99} \right) \Big/$

(245 391 497 880 823 860 888 299 579 254 871 888 794 227 262 599 280 ⋮
  227 626 878 561 171 419 833 316 545 174 421 298 255 912 601 373 ⋮
  297 161 029 158 231 721 510 947 939 098 833 986 781 088 403 218 ⋮
  180 280 686 067 327 394 838 801 692 178 191 801 614 949 332 339 ⋮
  280 285 444 598 672 020 587 366 832 067 114 230 479 977 376 797 ⋮
  699 352 031 451 207 170 874 710 206 026 698 758 582 371 695 524 ⋮
  670 940 358 191 728 591 918 945 312 500 000 000 000 000 000 000 ⋮
  000 000 000 000 000 000 000 000 000 000 000 000 000 000 000 000 ⋮
  000 000 000 000 000 000 000 000 000 000 000 000 000 000 000 000 ⋮
  000 000 000 000 000 000 000 000 000 000 000 000 000 000 000 000 ⋮
  000 000 000 000 000 000 000 000 000 000 000 000 000 000 000 000 ⋮
  000 000 000 000 000 000 000 000 000 000 000 000 000 000 000 000 ⋮
  000 000 000 000 000 000 000 000 000 000 000 000 000 000 000 000 ⋮
  000 000 000 000 000 000 000 000 000 000 000 000 000 000 000 000 ⋮
  000 000 000 000 000 000 000 000 000 000 000 000 000 000 a²⁰¹)

The integer in front of a^{201} is a very long number ($2.45... \times 10^{716}$) whose output has been iconized.

Figure 8.5 shows the radial probability for $n = 100$ and $\ell = 99$.

8.5   TRANSITIONS BETWEEN LEVELS

Consider the possible transition between two states of hydrogen. The expectation value of the dipole moment ($e\mathbf{r}$) sandwiched between the wave functions of the initial and final states,

$$\int \psi_f^*(-e\mathbf{r})\psi_i dV,$$

is a measure of whether or not the transition can occur. The semi-classical intuitive mechanism for the radiation is an oscillating electric dipole.

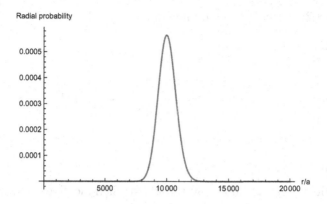

Figure 8.5   Radial probability for $n = 100$ and $\ell = 99$.

Example 8.24 Calculate the expectation value of the dipole vector connecting the $2p$ and $1s$ states.

```
In[33]:= ψf = ResourceFunction["HydrogenWavefunction"][{1, 0, 0},
             a, {r, θ, ϕ}];
         ψi = ResourceFunction["HydrogenWavefunction"][{2, 1, 0},
             a, {r, θ, ϕ}];
         p =
         - e
         ∫₀^∞ (∫₀^{2π} (∫₀^{π} r² Sin[θ] Conjugate[ψf]

                  FromSphericalCoordinates[{r, θ, ϕ}] ψi dθ) dϕ) dr
```

$$\text{Out[33]= } \left\{ 0\,e,\ 0\,e,\ a \left(-\frac{128\sqrt{2}}{243}\,e \right) \right\}$$

The lifetime (τ) of the state is proportional to the inverse square

$$\tau = \frac{\mu_0 \omega^3}{3\pi\hbar c} \left(\int \psi_f^*(-e\mathbf{r})\psi_i dV \right)^{-2},$$

where ω is the angular frequency of the photon emitted in the transition. It is the photon energy the transition divided by the power radiated from a classical oscillating electric dipole.

Example 8.25 Calculate the lifetime of the $2p \to 1s$ transition.

In[34]:= $\mathbf{E} = \dfrac{3\ e^{2}}{32\ \pi\ \varepsilon_{\theta}\ a_{\theta}}$; $\omega = \dfrac{\mathbf{E}}{\hbar}$;

$\mathtt{ScientificForm}\left[\mathtt{UnitConvert}\left[\left(\dfrac{\mu_{\theta}\ \omega^{3}}{3\ \pi\ \hbar\ c}\ \mathtt{p[\![3]\!]}^{2}\right)^{-1}\ \mathtt{/.\ a \to a_{\theta}}\right],\ 2\right]$

Out[35]//ScientificForm=

1.6×10^{-9} s

Electric dipole transitions are forbidden when the dipole integral is zero (corresponding to infinite lifetime). Transitions are also forbidden when the quantum number l does not change by one unit. This is a result of angular momentum conservation with the photon carrying one unit of angular momentum.

Example 8.26 Calculate the dipole factor for the $2s \to 1s$ transition ($\Delta\ell = 0$).

In[36]:= ψ_{i} = ResourceFunction["HydrogenWavefunction"][{2, 0, 0},

a, {r, θ, ϕ}];

$\mathtt{p} = \displaystyle\int_{\theta}^{\infty}\left(\int_{\theta}^{2\pi}\left(\int_{\theta}^{\pi}r^{2}\ \mathtt{Sin[\theta]\ Conjugate[\psi_{f}]} \times\right.\right.$

$\left.\left.\mathtt{FromSphericalCoordinates[\{r,\ \theta,\ \phi\}]}\ \psi_{i}\ \mathtt{d\theta}\right)\mathtt{d\phi}\right)\mathtt{d r}$

Out[36]= {0, 0, 0}

Example 8.27 Calculate the dipole factor for the $3d \to 1s$ transition ($\Delta\ell = 2$).

In[37]:= ψ_{i} = ResourceFunction["HydrogenWavefunction"][{3, 2, 0},

a, {r, θ, ϕ}];

$\mathtt{p} = \displaystyle\int_{\theta}^{\infty}\left(\int_{\theta}^{2\pi}\left(\int_{\theta}^{\pi}r^{2}\ \mathtt{Sin[\theta]\ Conjugate[\psi_{f}]} \times\right.\right.$

$\left.\left.\mathtt{FromSphericalCoordinates[\{r,\ \theta,\ \phi\}]}\ \psi_{i}\ \mathtt{d\theta}\right)\mathtt{d\phi}\right)\mathtt{d r}$

Out[37]= {0, 0, 0}

8.6 ELECTRON INTRINSIC ANGULAR MOMENTUM

There is one more quantum number for an electron in the hydrogen atom, the physics of which is not contained in the Schrödinger equation. The electron has an intrinsic angular momentum, commonly referred to as spin, which can take on 2 possible values. The spin quantum number (s) can be $\frac{1}{2}$ or $-\frac{1}{2}$. This corresponds to a contribution to the angular momentum (S) of

$$S = \sqrt{s(s+1)}\hbar = \frac{\sqrt{3}}{2}\hbar.$$

The total angular momentum of the electron ($\mathbf{J}$) is the vector addition

$$\mathbf{J} = \mathbf{L} + \mathbf{S}.$$

8.6.1 Addition of Angular Momentum

The rules for adding angular momentum quantum numbers is that the sum can take on all values from $|\ell - s|$ to $\ell + s|$, Thus,

$$0 + \frac{1}{2} \rightarrow \frac{1}{2},$$

and

$$1 + \frac{1}{2} \rightarrow \frac{1}{2} \text{ or } \frac{3}{2}.$$

For the $n = 2$ levels of hydrogen these states are labeled $1s_{1/2}, 2p_{1/2}, 2p_{3/2}$, where the subscript denotes the value of the quantum number j and

$$J = \sqrt{j(j+1)}\hbar.$$

8.6.2 Magnetic Moment

Compared to classical orbital motion of a charge whose magnetic moment is

$$\mu = \frac{q}{2m}\mathbf{L},$$

the electron has a gyromagnetic factor (g) that accounts for the spin quantum number being 1/2,

$$\mu = g\frac{e}{2m}\mathbf{S}.$$

The value of g for the electron is very nearly equal to -2. (Note that the electron's negative charge is contained in g when it is written this way.) With

$g = -2$, 1/2 unit of intrinsic angular momentum gives the same magnetic contribution as one unit of orbital angular momentum.

The Bohr magneton is the dipole strength of the electron's permanent magnet. It is the projection of the magnetic moment in some direction ($S \rightarrow s$) and defined with $g = 2$ so that the spin $\frac{1}{2}$ factor exactly cancels, giving

$$\mu_B = \frac{e\hbar}{2m}$$

Example 8.28 Get the Bohr magneton.

In[38]:= `Quantity["BohrMagneton"]`

Out[38]=

μ_B

Example 8.29 Get the value of the Bohr magneton.

In[39]:= `NumberForm[UnitConvert[`μ_B`], 3]`

Out[39]//NumberForm=

9.27×10^{-24} m^{2}A

8.6.3 Zeeman Effect

A magnet has lower (higher) energy when aligned (anti-aligned) with an external magnetic field, giving an energy shift of

$$\Delta E = \mu \cdot \mathbf{B} = \mu_z B_z = \frac{e}{2m}(L_z + 2S_z)B_z = \frac{e\hbar}{2m}(m_\ell + 2m_s)B_z$$

compared to zero field. Here $\mathbf{B} = B_z \hat{z}$.

Example 8.30 Calculate the magnetic field need to make a fractional energy shift of 10^{-4} on the $3d_{5/2}$ state of hydrogen.

In[40]:= `l = 2; s = `$\frac{1}{2}$`; `μ` = `μ_B` (l + 2 s); B = UnitConvert[`$\frac{1}{\mu}$` 10`$^{-4}$` `$\frac{13.6 \text{ eV}}{9}$`, T]`

Out[40]=

0.870198 T

8.6.4 Spin-Orbit Interaction

From the viewpoint of the electron in a p ($\ell = 1$) state, the orbiting proton makes a magnetic field with which the electron's magnet moment interacts, causing an energy shift between the $2p_{1/2}$ and $2p_{3/2}$ states, which are otherwise degenerate in the solution of Schrödinger's equation (8.2). The magnetic field (**B**) may be estimated as that at the center of of a tiny current loop with radius (r). From Ampère's law,

$$\mathbf{B} = \frac{\mu_0}{4\pi} \int \frac{(Id\mathbf{l}) \times \hat{\mathbf{r}}}{r^2} = \frac{\mu_0}{4\pi} \frac{2\pi r}{r^2} = \frac{\mu_0 I}{2r},$$

where μ_0 is the magnetic constant and I is the current from a proton charge with orbit frequency ($\frac{v}{2\pi r}$),

$$I = \frac{ev}{2\pi r}.$$

This gives

$$B = \frac{\mu_0 ev}{4\pi r^2}.$$

Example 8.31 Get the magnetic constant.

In[41]:= `Quantity["MagneticConstant"]`

Out[41]=

μ_0

The magnetic constant, is related to the electric constant by

$$\mu_0 = \frac{1}{c^2 \varepsilon_0}.$$

Example 8.32 Check the relationship between the electric and magnetic constants.

In[42]:= μ_0 `==` $\dfrac{1}{\varepsilon_0 \, c^2}$

Out[42]=

True

The energy shift caused by the electron magnetic moment is

$$\Delta E = \mu \cdot \mathbf{B} = g \frac{e}{2m} \mathbf{S} \cdot \mathbf{B} = 2 \left(\frac{e}{2m} \right) \left(\frac{\hbar}{2} \right) \left(\frac{\mu_0 ev}{4\pi r^2} \right).$$

For the first excited state in the Bohr model (5.2.2),

$$v = \frac{c\alpha}{2},$$

and

$$r = \frac{4\hbar}{\alpha mc}.$$

This gives

$$\Delta E = \frac{\alpha^4 mc^2}{64}.$$

Example 8.33 Solve for ΔE.

In[43]:= $v = \dfrac{c\,\alpha}{2}$; $r = \dfrac{4\,\hbar}{\alpha\,m\,c}$;

$\Delta E = 2\,\dfrac{e}{2\,m}\,\dfrac{\hbar}{2}\,\dfrac{\mu_\theta\,e\,v}{4\,\pi\,r^2}\;/.\;\left\{\mu_\theta \to \dfrac{1}{c^2\,\varepsilon_0}\,,\;\hbar \to \dfrac{e^2}{4\,\pi\,\varepsilon_0\,\alpha\,c}\right\}$

Out[44]=

$\dfrac{1}{64}\,c^2\,m\,\alpha^4$

In solving for ΔE in Ex. 8.33, the input quantities are user-defined variables (not units) to allow the substitutions $\mu_0 \to \frac{1}{c^2\varepsilon_0}$ and $\hbar \to \frac{e^2}{4\pi\varepsilon_0 c}$. The energy levels are themselves proportional to α^2 and the correction to them is proportional to α^4. This is how α got its name as the *fine structure constant*. Note that the correction to spin up (down) is positive (negative) so the energy difference between the $2p_{3/2}$ and $2p_{1/2}$ states is $2\Delta E$.

Example 8.34 Calculate the energy difference between the $2p_{3/2}$ and $2p_{1/2}$ states.

In[45]:= $m = \boxed{\text{electron}\;\;\text{PARTICLE}}\left[\boxed{\textit{mass}}\right];$

$\texttt{ScientificForm}\left[\texttt{UnitConvert}\left[2 \times \dfrac{1}{64}\,\alpha^4\,m\,c^2,\;\text{eV}\right],\,2\right]$

Out[46]//ScientificForm=

4.5×10^{-5} eV

Despite using a simple Bohr orbit for getting the magnetic field, the energy splitting turns out to be correct. Such semi-classical classical calculations are often extremely physically intuitive compared to a full calculation with wave functions.

8.6.5 Hyperfine Splitting

An electron in the ground hydrogen atom has zero orbital angular momentum, however, it will still feel the magnetic field caused by the internal magnet of the proton (due to intrinsic angular momentum). There will be an energy difference between an electron having spin up or spin down (Fig. 8.6).

The energy difference between spin orientations may be estimated by evaluating the magnetic field (Fig. 8.7) from the proton at the location of the electron. The proton magnetic moment is

$$\mu_\text{p} = g_\text{p}\dfrac{e\hbar}{2m_\text{p}},$$

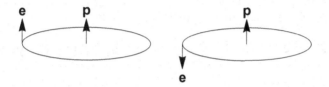

Figure 8.6   Visualization of the spin-flip transition in hydrogen. The state with spins aligned is higher energy than spins opposite.

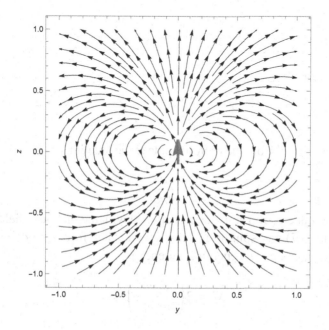

Figure 8.7   Dipole field from the proton magnetic moment.

where analogous to the electron case, g_p is the proton g-factor to account for the charge-to-mass ratio and m_p is the proton mass.

The expression for the field can be calculated from

$$\mathbf{A} = \frac{\mu_0}{4\pi}\mu_\mathrm{p}\nabla\times\frac{\hat{\mathbf{z}}}{r},$$

and

$$\mathbf{B} = \nabla\times\mathbf{A}.$$

Example 8.35 Calculate the magnetic dipole field.

$$\text{In[47]:= } A = \frac{\mu_\theta}{4\pi}\,\mu_p\,\text{Curl}\left[\frac{1}{\sqrt{x^2 + y^2 + z^2}}\,\{0,\,0,\,1\},\,\{x,\,y,\,z\}\right];$$

$$B = \text{Curl}[A,\,\{x,\,y,\,z\}]$$

$$\text{Out[48]= } \left\{\frac{3\,x\,z\,\left(3\,\mu_{B\,\theta}\right)\,\left(3\,\mu_{B\,\{0,0,0\}}\right)}{4\pi\left(x^2 + y^2 + z^2\right)^{5/2}},\right.$$

$$\frac{3\,y\,z\,\left(3\,\mu_{B\,\theta}\right)\,\left(3\,\mu_{B\,\{0,0,0\}}\right)}{4\pi\left(x^2 + y^2 + z^2\right)^{5/2}},\,-\frac{3\,x^2\,\left(3\,\mu_{B\,\theta}\right)\,\left(3\,\mu_{B\,\{0,0,0\}}\right)}{4\pi\left(x^2 + y^2 + z^2\right)^{5/2}}-$$

$$\left.\frac{3\,y^2\,\left(3\,\mu_{B\,\theta}\right)\,\left(3\,\mu_{B\,\{0,0,0\}}\right)}{4\pi\left(x^2 + y^2 + z^2\right)^{5/2}}+\frac{\left(3\,\mu_{B\,\theta}\right)\,\left(3\,\mu_{B\,\{0,0,0\}}\right)}{2\pi\left(x^2 + y^2 + z^2\right)^{3/2}}\right\}$$

The magnetic field from the proton dipole at the location of the electron (Bohr radius, a) is

$$B = \frac{\mu_0}{4\pi a^3}\mu_p.$$

Notice that electron spin down is aligned with the magnetic field from the proton (Fig. 8.7) and spin up is anti-aligned. Therefore, spin down is lower energy than spin up. The energy difference is

$$\Delta E = 2\left(\frac{\mu_0 e\hbar}{4\pi a^3 m_p}\right)g_p g_e$$

Example 8.36 Calculate the difference in energy levels for electron and proton spins aligned or anti-aligned.

$$\text{In[50]:= } B = \frac{\mu_\theta}{4\pi\,{a_\theta}^3}\,\frac{e\,\hbar}{2\,\boxed{\text{proton} \text{ PARTICLE}}\left[\boxed{mass}\right]}$$

$$\boxed{\text{proton} \text{ PARTICLE}}\left[\boxed{\text{spin g-factor}}\right];$$

$$N\left[\text{UnitConvert}\left[2\left(-\boxed{\text{electron} \text{ PARTICLE}}\left[\boxed{\text{spin g-factor}}\right]\right)\mu_B\,B,\,eV\right],\right.$$

$$\left.1\right]$$

$$\text{Out[50]= } 4.\times 10^{-6}\ eV$$

This is the correct order of magnitude for the observed energy difference. To get a more accurate answer, on needs to use the $1s$ hydrogen wave function (8.1.2) and evaluate its square over the magnetic field produced by the proton. The only contribution is at the origin where the magnetic field is a delta function. The magnetic field at the proton location is

$$\mathbf{B} = \frac{2}{3}\mu\delta(\mathbf{r}).$$

The delta function has the property that its value is zero everywhere except at the origin where it is infinite, such that its integral gives unity. It can be thought of as a Gaussian with unit area under the curve in the limit where its width goes to zero.

Example 8.37 Calculate the energy difference using the wave function.

In[51]:= ΔE = N[UnitConvert[$\frac{4}{3}\frac{\mu_\theta}{2\pi\,a_\theta{}^3}$ ×

$\mu_B\ \dfrac{e\,\hbar}{2\ \boxed{\text{\textbf{proton} \small PARTICLE}}\ \boxed{\boxed{\text{\textit{mass}}}}}\ \boxed{\text{\textbf{proton} \small PARTICLE}}\left[\boxed{\text{\textit{spin g-factor}}}\right]$

$\left(-\boxed{\text{\textbf{electron} \small PARTICLE}}\left[\boxed{\text{\textit{spin g-factor}}}\right]\right),\ \text{eV}\right],\ 3]$

Out[51]= 5.88×10^{-6} eV

This answer is correct to order α.

Example 8.38 Calculate the wavelength of the radiation emitted when the electron flips its spin.

In[52]:= N[UnitConvert[$\frac{h\ c}{\Delta E}$, cm], 3]

Out[52]=
$\qquad$ 21.1 cm

This is the famous 21-cm line from hydrogen that allowed the mapping of the spiral nature of the Milky Way galaxy due to its ability to penetrate interstellar dust.

The lifetime of the hyperfine transition can be calculated from the classical formula for radiated power,

$$P = \mu_0\frac{\mu_p^2\omega^4}{12\pi c}.$$

The lifetime is

$$\tau = \frac{\Delta E}{P}.$$

Example 8.39 Calculate the lifetime of the hyperfine transition.

In[53]:= μ_p = - $\boxed{\text{electron} \quad \text{PARTICLE}}\left[\boxed{\text{spin g-factor}}\right] \mu_B$; $\omega = \dfrac{\Delta E}{\hbar}$;

$$P = \frac{\mu_\theta \, \mu_p^2 \, \omega^4}{12 \, \pi \, c^3} \, ;$$

ScientificForm$\left[\text{UnitConvert}\left[\dfrac{\Delta E}{P}, \, \text{yr}\right], \, 2\right]$

Out[55]//ScientificForm=

$1.1 \times 10^7 \, \text{yr}$

8.6.6 Lamb Shift

In the solution to the Schrödinger equation, even after accounting for electron spin and spin-orbit interaction, the $2s_{1/2}$ and $2p_{1/2}$ states have identical energy (8.2.2). The wave functions (Figs. 8.2 and 8.3) of these 2 states, however, are very different. The electron has a self-interaction in which it can emit and absorb virtual photons. (A virtual photon is one that never enters the real world, but can "exist" on borrowed energy for a short time allowed by the uncertainty principle.) The effect of the self-interaction spreads out the electron on a distance scale of about 0.1 fm. The $2s$ wave function has a little piece that brings the electron close to the nucleus. Without the electron self-interaction, the energies are identical, but when it is included, the $2s_{1/2}$ energy becomes slightly higher because the smearing results in the electron feeling a smaller force. A smaller force translates to less binding energy and a larger total energy.

The energy difference between the $2s_{1/2}$ and $2p_{1/2}$ states is called the Lamb shift after Willis lamb who first measured the energy difference by measuring the transitions

$$2s_{1/2} \rightarrow 2p_{1/2} \rightarrow 1s_{1/2}.$$

(Note that the direct transition $2s \rightarrow 1s$ is forbidden.) The frequency of the photons in the transition $2s_{1/2} \rightarrow 2p_{1/2}$ is 1058 Hz. The Lamb shift is the same order of magnitude as the hyperfine splitting (8.6.5) and is proportional to the wave function squared at the origin.

Example 8.40 Calculate the wave function squared at the origin for the $2s_{1/2}$ and $2p_{1/2}$ states.

```
In[56]:= ClearAll["Global`*"];
        ψ = ResourceFunction["HydrogenWavefunction"][{2, 0, 0},
          a, {r, θ, ϕ}];
        ψ² /. r → 0
        ψ = ResourceFunction["HydrogenWavefunction"][{2, 1, 0},
          a, {r, θ, ϕ}];
        ψ² /. r → 0 // FullSimplify
```

$$\text{Out[57]=} \quad \frac{1}{8\,a^3\,\pi}$$

Out[59]= 0

There is no shift for the $2p_{1/2}$ state. The $2s_{1/2}$ state is shifted up by an amount proportional to

$$\left|\psi_{2s}(0)^2\right| = \frac{1}{8\pi a^3} = \frac{(mc^2)^3}{8\pi(4\pi\hbar^2\varepsilon_0)^3}.$$

The expression to leading order is,

$$\Delta E = \alpha^5 \frac{mc^2}{6\pi} \ln\left(\frac{1}{8.9\alpha^2}\right).$$

Example 8.41 Calculate the Lamb shift and transition frequency.

```
In[60]:= m = [ electron PARTICLE ] [ mass ];
        ScientificForm[UnitConvert[
```

$$\Delta E = \alpha^5 \frac{m\ c^2}{6\ \pi} \text{Log}\left[\text{UnitConvert}\left[\frac{1}{8.9\ \alpha^2}\right]\right], \text{ eV}], 4]$$

```
        NumberForm[UnitConvert[
```
$\frac{\Delta E}{h}$, MHz], 4]

Out[61]//ScientificForm=

4.294×10^{-6} eV

Out[62]//NumberForm=

1038. MHz

This result is good to about 2%.

Example 8.42 Use the measured value of the $2s_{1/2}$ Lamb shift to estimate the Lamb shift frequency for the $3s_{1/2}$ state.

```
In[63]:= ψ = ResourceFunction["HydrogenWavefunction"][{3, 0, 0},
            a, {r, θ, ϕ}];
         P1 = ψ² /. r → 0 ;
         ψ = ResourceFunction["HydrogenWavefunction"][
            {2, 0, 0}, a, {r, θ, ϕ}];
         P2 = ψ² /. r → 0 ;
         N[1058 MHz P1/P2, 3]
```

Out[67]= 313. MHz

The measurement of the Lamb shift and the calculation that followed ushered in the development of quantum electrodynamics which is the foundation on which the standard model of particle physics was constructed.

8.6.7 The Electron g-Factor

Due to the self-interaction of the electron, the spin component ($g/2$) contributes slightly more to the total magnetic moment than one unit of orbital angular momentum. If the g-factor were exactly 2, there would be no difference in the energies of the states. The self-interaction alters the value of g by about 0.1% (8.6.2). Thus, the electron in a $2s_{1/2}$ state is a slightly stronger magnet than in the $2p_{1/2}$ state.

Example 8.43 Get the g-factor of the electron.

```
In[68]:= UnitConvert[ electron PARTICLE [ spin g-factor ]]
```

Out[68]= -2.0023193043622

The g-factor of the electron has been calculated to 10 decimal places (order α^5) using the theory of quantum electrodynamics. The leading corrections are

$$\frac{|g|-2}{2} = \frac{\alpha}{2\pi} + \left(\frac{\alpha}{2\pi}\right)^2 \left(\frac{197}{36} + \frac{\pi^2}{3} - 2\pi^2 \ln 2 + 3\zeta(3)\right),$$

where

$$\zeta(3) = \frac{1}{2} \int_0^\infty \frac{x^2}{e^x - 1} \approx 1.20206.$$

Example 8.44 Calculate $\frac{|g|-2}{2}$ to order α^2.

In[69]:= `N[UnitConvert[`$\frac{\alpha}{2\pi} + \left(\frac{\alpha}{\pi}\right)^2 \left(\frac{197}{144} + \frac{\pi^2}{12} - \frac{\pi^2}{2}\right.$ `Log[2]` $\left.+ \frac{3\,\text{Zeta}[3]}{4}\right)$`], 5]`

Out[69]=

0.0011596

CHAPTER 9

Statistical Physics

9.1 PROBABILITY DISTRIBUTIONS

Probability distributions are at the heart of modern physics and have been encountered in thermal radiation (2.2), radioactive decay (3.1) as well as $|\psi^2|$ from the solution to the Schrödinger equation (6.5). In general, a probability distribution may be written as $P(x)$ which represents some probability per unit of x. The function is usually normalized to unity,

$$\int P(x)dx = 1.$$

9.1.1 Binomial Distribution

Some distributions like flipping a coin are not continuous but have discreet outcomes. The binomial distribution $f_b(x)$ represents the probability of getting an integer result x from n trials when the probability per trial is p,

$$f_b(x) = \frac{n!p^x(1-p)^{n-x}}{x!(n-x)!}.$$

The normalization condition is

$$\sum_{x=0}^{n} f_b(x) = 1.$$

Example 9.1 Sum all the binomial terms.

In[1]:= $\sum_{x=0}^{n} \frac{(n! \, p^x \, (1-p)^{n-x})}{x! \, (n-x)!}$

Out[1]= 1

DOI: 10.1201/9781003395515-9

Example 9.2 Calculate the probability of getting a coin toss in your favor exactly 5 times out of 10 tosses, assuming the probability of the coin landing either way is 1/2.

In[2]:= $\dfrac{n! \, p^x \, (1 - p)^{n-x}}{x! \, (n - x)!}$ /. $\{n \to 10, p \to .5, x \to 5\}$

Out[2]= 0.246094

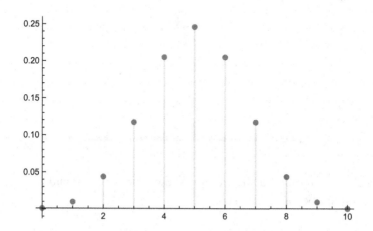

Figure 9.1   The binomial distribution is shown for $n = 10$ and $p = 1/2$.

The average is
$$\langle x \rangle = np,$$
and the rms is
$$x_{rms} = \sqrt{np(1 - p)}.$$

Example 9.3 Calculate the average of a binomial distribution.

In[3]:= $\displaystyle\sum_{x=0}^{n} \dfrac{x \, \left(n! \, p^x \, (1 - p)^{n-x}\right)}{x! \, (n - x)!}$

Out[3]= $n \, p$

Example 9.4 Calculate the rms of a binomial distribution.

In[4]:= $\displaystyle\sqrt{\sum_{x=0}^{n} \dfrac{(x - n \, p)^2 \, \left(n! \, p^x \, (1 - p)^{n-x}\right)}{x! \, (n - x)!}}$

Out[4]= $\sqrt{-n \left(-p + p^2\right)}$

In the limit where $n \to \infty$, the variable x becomes continuous and the binomial distribution because a Gaussian, provided that the average np is not too small. In the case where $p << 1$, one gets a Poisson distribution.

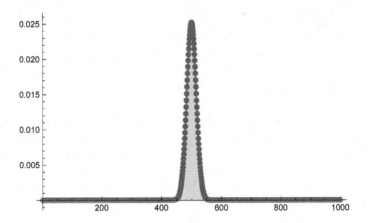

Figure 9.2   The binomial distribution is shown for $n = 1000$ and $p = 1/2$. The distribution has become Gaussian.

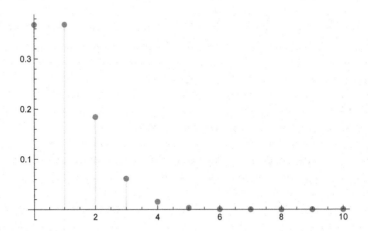

Figure 9.3   The binomial distribution is shown for $n = 1000$ and $p = 0.001$. The distribution has become Poisson.

9.1.2 Poisson Distribution

The Poisson distribution corresponds to large n and small x and is obtained from the binomial distribution with the approximations

$$(1-p)^n = e^{n\ln(1-p)} \approx e^{-np},$$

and

$$\frac{n!}{(n-x)!} = (n)(n-1)(n-2)...(n-x) \approx n^x.$$

With both $p \ll 1$ and $x \ll n$, the binomial distribution reduces to the Poisson distribution

$$f_P(x) = \frac{e^{-a}a^x}{x!},$$

where $a = np$, the average.

Example 9.5 Expand $\ln(1-p)$ about $p = 0$.

In[5]:= **Series[Log[1-p], {p, 0, 1}]**

Out[5]= $-p + O[p]^2$

The Poisson distribution is normalized to unity.

Example 9.6 Sum all the Poisson terms.

In[6]:= $\sum_{x=0}^{\infty} \frac{e^{-a} a^x}{x!}$

Out[6]= 1

A major-use case of the Poisson distribution comes about when one wants to know an upper limit on the average value (a) when a small number of occurrences are observed. For example, suppose 0 events are observed. What is the upper limit on a at 90% confidence level (CL)? The answer is given by

$$f_P(0) = \frac{e^{-a}a^0}{0!} = e^{-a} = 0.1.$$

Solving for a,

$$a = -\ln 0.05 = 2.3.$$

One says that a is less than 2.3 at 90% CL.

Example 9.7 Find the 90% CL limits for 0 to 3 events observed.

In[7]:= `$Assumptions = a > 0;`

$$\text{Table}\left[\text{Solve}\left[\sum_{x=0}^{i} \frac{e^{-a}\, a^{x}}{x!} == 0.1, a\right], \{i, 0, 3\}\right]$$

Out[8]= `{{{a → 2.30259}}, {{a → 3.88972}}, {{a → 5.32232}}, {{a → 6.68078}}}`

Example 9.8 Find the 95% CL limits for 0 to 3 events observed.

In[9]:= `$Assumptions = a > 0;`

$$\text{Table}\left[\text{Solve}\left[\sum_{x=0}^{i} \frac{e^{-a}\, a^{x}}{x!} == 0.05, a\right], \{i, 0, 3\}\right]$$

Out[10]=

`{{{a → 2.99573}}, {{a → 4.74386}}, {{a → 6.29579}}, {{a → 7.75366}}}`

9.1.3 Gaussian Distribution

The Gaussian distribution is the result of the sum of random processes. As pointed out, it is obtained from the binomial distribution in the limit of large n and np that is not small. Figure 9.4 shows a histogram of the sum 10 random numbers between 0 and 1, done 10^{6} times. The result is a Gaussian.

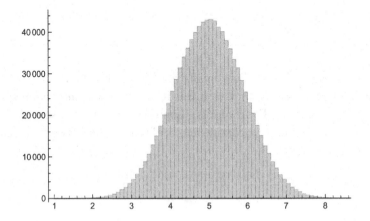

Figure 9.4 For a histogram of the sum of 10 random numbers between 0 and 1, the average is 5.00 and the rms is 0.913.

The Gaussian distribution may be written

$$f_G = Ce^{-(x-a)^2/2\sigma^2},$$

where a is the average, σ is the standard deviation, and C is the normalization constant. When the average is zero,

$$f_G = Ce^{-x^2/2\sigma^2},$$

Example 9.9 Calculate the normalization constant.

In[11]:= **\$Assumptions** = σ > 0; C = $\left(\int_{-\infty}^{\infty} e^{(-x^2)/(2\sigma^2)} \, dx \right)^{-1}$

Out[11]=

$$\frac{1}{\sqrt{2\pi}\,\sigma}$$

Example 9.10 Calculate the Gaussian rms.

In[12]:= $\sqrt{\dfrac{1}{\sqrt{2\pi}\,\sigma} \int_{-\infty}^{\infty} x^2\, e^{(-x^2)/(2\sigma^2)} \, dx}$ **// Simplify**

Out[12]=

σ

About 32% of the distribution is outside 1 σ (on either side).

Example 9.11 Calculate the probability to exceed n sigma (on either side), for n from 1 to 5.

In[13]:= **Table**$\left[\text{ScientificForm}\left[2\,\text{NIntegrate}\left[\dfrac{1}{\sqrt{2\pi}}\, e^{-x^2/2}, \{x, i, \infty\} \right], 3 \right], \{i, 1, 5\} \right]$

Out[13]=

$\{3.17 \times 10^{-1}, 4.55 \times 10^{-2}, 2.7 \times 10^{-3}, 6.33 \times 10^{-5}, 5.73 \times 10^{-7}\}$

9.2 MAXWELL-BOLTZMANN DISTRIBUTION

The Maxwell-Boltzmann (MB) distribution applies to a large number of particles that are in thermal equilibrium and are distinguishable because they are far apart and their wave properties do not overlap. It is the classical regime.

9.2.1 The Ideal Gas

Temperature on an absolute scale is a measure of average kinetic energy. For an ideal gas, that scale may be defined as

$$\langle K \rangle \equiv \frac{3}{2}kT = \frac{1}{2}m\langle v^2 \rangle.$$

Thus, temperature is also a measure of the rms speed.

Example 9.12 Calculate the rms speed of nitrogen molecules at $T = 300$ K.

In[14]:= $\text{N}\left[\text{UnitConvert}\left[\sqrt{\dfrac{3\ \text{k}\ \text{T}}{\text{m}}}\ /.\ \left\{\text{T} \to 300\ \text{K}\ ,\ \text{m} \to \boxed{\text{nitrogen}\ \text{CHEMICAL}}\left[\boxed{molecular\ mass}\right]\right\}\right],\ 3\right]$

Out[14]=

    517. m/s

9.2.2 Gas Pressure

The gas pressure is proportional to the average speed squared. Consider gas molecules hitting a walls of a container. The pressure gets one power of v from momentum transfer during a collision and another power of v for the probability per time of hitting the wall. This gives $P \sim v^2 \sim T$ and is written as the ideal gas law,

$$P = \frac{N}{V}kT.$$

Example 9.13 Calculate the volume of 1 mole (Avogadro's number) of molecules at standard temperature and pressure (STP) of 273 K and 1 atm.

In[15]:= $\text{V} = \text{N}\left[\text{UnitConvert}\left[\dfrac{\text{N}_\theta\ \text{k}\ \text{T}}{\text{P}}\right]\ /.\ \left\{\text{T} \to 273\ \text{K}\ ,\ \text{P} \to 1\ \text{atm}\right\}\right]$

Out[15]=

    0.0224017 m^3

9.2.3 Mean Free Path

A cylinder traced out by a molecule as it moves between collisions may be equated to the volume that contains one molecule (see Fig. 9.5),

$$\pi a^2 l \approx d^3,$$

where a is the molecular size, d is the average distance between molecules, and l is the mean free path between collisions. This gives

$$l = \frac{d^3}{\pi a^2}.$$

Writing $n = d^{-3}$ is the number of particles per volume,

$$l = \frac{1}{n\sigma},$$

where $\sigma = \pi a^2$ as the collision cross section as defined in 3.6.3. The mean free path is inversely proportional to both the number of particles per volume and the collision cross section.

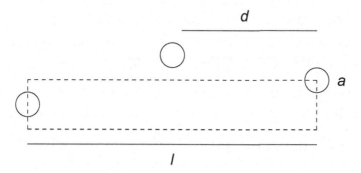

Figure 9.5 Schematic to indicate the mean free path (l) and its qualitative relationship to molecular size (a) and average distance between molecules (d).

Example 9.14 Estimate the mean free path of a molecule at STP.

In[16]:= $\quad$ **a = 0.3 nm ; ScientificForm$\left[\text{UnitConvert}\left[\dfrac{V}{N_\theta\,\pi\,a^2}\right], 1\right]$**

Out[16]//ScientificForm=

$\quad$ $1. \times 10^{-7}$ m

Example 9.15 Estimate the average distance between molecules at STP.

In[17]:= $\quad$ **ScientificForm$\left[\text{UnitConvert}\left[V\,/\,N_\theta\,\right]^{1/3}, 1\right]$**

Out[17]//ScientificForm=

$\quad$ $3. \times 10^{-9}$ m

The average kinetic energy may be written

$$\frac{\langle p^2 \rangle}{2m} = \frac{3}{2}kT.$$

Therefore, the rms momentum is

$$\sqrt{\langle p^2 \rangle} = \sqrt{3kT}.$$

Example 9.16 Calculate the de Broglie wavelength of nitrogen molecules at STP.

In[18]:= m = [**nitrogen** CHEMICAL] [*molecular mass*] ;

T = 273 K ;

ScientificForm[UnitConvert[$\dfrac{h}{\sqrt{3\,m\,k\,T}}$], 1]

Out[20]//ScientificForm=

$$3. \times 10^{-11} \text{ m}$$

The idea gas is characterized by

$$a \ll d \ll l.$$

9.2.4 Velocity Distribution

The velocity distribution for each of the x, y, x components is Gaussian as a result of random collisions and may be written

$$f(v_x) = \sqrt{\frac{m}{2\pi kT}} e^{-mv_x^2/2kT},$$

$$f(v_y) = \sqrt{\frac{m}{2\pi kT}} e^{-mv_y^2/2kT},$$

$$f(v_z) = \sqrt{\frac{m}{2\pi kT}} e^{-mv_z^2/2kT}.$$

Each component is symmetric about zero and has zero average.

Example 9.17 Check the normalization of the velocity distribution.

In[21]:= $Assumptions = {m > 0, k > 0, T > 0};

$$\int_{-\infty}^{\infty} \sqrt{\frac{m}{2\pi kT}} \ e^{-\frac{mv^2}{2kT}} dv$$

Out[22]=

1.0000

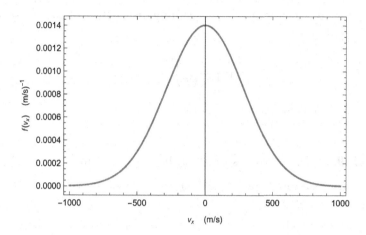

Figure 9.6 The Maxwell-Boltzmann velocity distribution is shown for nitrogen molecules at STP.

Example 9.18 Calculate the rms of the velocity distribution.

In[23]:= **Clear[m, k, T]; $Assumptions = {m > 0, k > 0, T > 0};**

$$\sqrt{\int_{-\infty}^{\infty} v^2 \sqrt{\frac{m}{2\pi k T}}\ e^{-\frac{m v^2}{2 k T}}\, dv}$$

Out[24]=

$$\sqrt{\frac{k T}{m}}$$

It is seen that

$$\frac{1}{2}m\langle v_x^2\rangle = \frac{1}{2}m\langle v_y^2\rangle = \frac{1}{2}m\langle v_z^2\rangle = \frac{1}{2}kT.$$

Each component of the velocity contributes $\frac{1}{2}kT$ to the kinetic energy. This result is known as the equipartition theorem. Together, the 3 translational components contribute $\frac{3}{2}kT$ to the kinetic energy in agreement with the definition of temperature (9.2.1).

9.2.5 Speed Distribution

The speed distribution depends only on the magnitude, not direction. The speed distribution is obtained by integrating over all angles,

$$dv_x dv_y dv_z = v^2 \sin\theta\, dv\, d\theta\, d\phi,$$

which produces a factor of v^2, The resulting distribution of particle speeds in thermal equilibrium is

$$f(v) = 4\pi \left(\frac{m}{2\pi kT}\right)^{3/2} v^2 e^{-mv^2/2kT}$$

The distribution is normalized to unity.

Example 9.19 Integrate the speed distribution from 0 to ∞.

In[25]:= **$Assumptions = {m > 0, k T > 0};**

$$\int_0^\infty 4\,\pi \left(\frac{m}{2\,\pi\,k\,T}\right)^{3/2} v^2\,e^{-\frac{m v^2}{2 k T}}\,dv$$

Out[26]=

1

The most probable speed is found by setting the derivative of the distribution equal to 0.

Example 9.20 Calculate the most probable speed.

In[27]:= **$Assumptions = {v > 0, m > 0, k > 0, T > 0};**

Solve$\left[D\left[v^2\,e^{-\frac{m v^2}{2 k T}},\,v\right] == 0\right]$ // Simplify

Out[28]=

$$\left\{\left\{v \to \sqrt{2}\,\sqrt{\frac{k\,T}{m}}\right\}\right\}$$

The dimensionless speed distribution $N(x)$ in units of most probable speed is

$$N(x) = \frac{4}{\sqrt{\pi}} x^2 e^{-x}$$

Example 9.21 Calculate fraction of particles that are within 10% of the most probable speed.

In[29]:= **NIntegrate$\left[\frac{4}{\sqrt{\pi}}\,x^2\,e^{-x^2},\,\{x,\,.9,\,1.1\}\right]$**

Out[29]=

0.164941

The average speed is calculated from

$$v_{\text{ave}} = \int_0^\infty v f(v) dv$$

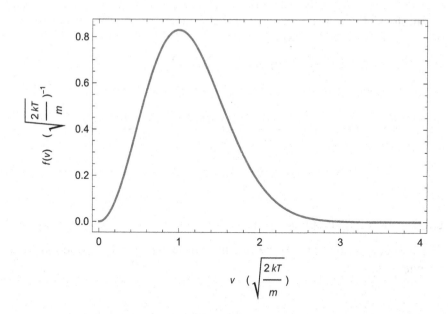

Figure 9.7 The Maxwell speed distribution is shown in units of the most probable speed.

Example 9.22 Calculate the average speed.

In[31]:= $\text{Integrate}\left[4\pi\left(\dfrac{m}{2\pi kT}\right)^{3/2} v^3\, e^{-\frac{mv^2}{2kT}},\ \{v,\ 0,\ \infty\}\right]$

Out[31]=

$$2\sqrt{\dfrac{2}{\pi}}\ \sqrt{\dfrac{kT}{m}}$$

The root-mean-square (rms) speed is calculated from

$$v_{\text{rms}} = \sqrt{\int_0^\infty v^2 f(v)\,dv}$$

Example 9.23 Calculate the rms speed.

In[32]:= $\sqrt{\text{Integrate}\left[4\pi\left(\dfrac{m}{2\pi kT}\right)^{3/2} v^4\, e^{-\frac{mv^2}{2kT}},\ \{v,\ 0,\ \infty\}\right]}$

Out[32]=

$$\sqrt{3}\ \sqrt{\dfrac{kT}{m}}$$

Example 9.24 Calculate the most probable, average, and rms speeds for nitrogen molecules at $T = 300$ K.

In[33]:= **m =** [**nitrogen** CHEMICAL] [*molecular mass*] **;**

T = 300 K ;

N [UnitConvert [{ $\sqrt{\dfrac{2 \text{ k T}}{m}}$, $\sqrt{\dfrac{8 \text{ k T}}{\pi \text{ m}}}$, $\sqrt{\dfrac{3 \text{ k T}}{m}}$ }] **, 3**]

Out[35]=

$\{ 422. \text{ m/s} , 476. \text{ m/s} , 517. \text{ m/s} \}$

9.2.6 Energy Distribution

The energy distribution may be obtained from the speed distribution by a change in variables,

$$E = \frac{1}{2} m v^2,$$

where E is the kinetic energy. The differential is

$$dE = mv\,dv,$$

and

$$v^2 dv = \frac{v}{m} dE = \frac{1}{m} \sqrt{\frac{2E}{m}} dE.$$

Thus, the energy distribution goes as $\sqrt{E} e^{-E/kT}$. The normalized distribution is

$$f(E) = \frac{2}{\sqrt{\pi}} (kT)^{-3/2} \sqrt{E} e^{-E/kT}$$

Example 9.25 Verify that the energy distribution is normalized to unity.

In[36]:= $\dfrac{1}{\sqrt{\pi}} \displaystyle\int_0^\infty 2 \, (\text{k T})^{-3/2} \, E^{1/2} \, e^{-\frac{E}{kT}} \, dE$

Out[36]=

1

Example 9.26 Calculate average energy.

In[37]:= **Clear[k, T]; $Assumptions = {T > 0, k > 0};**

$\dfrac{1}{\sqrt{\pi}} \displaystyle\int_0^\infty 2 \, (\text{k T})^{-3/2} \, E^{3/2} \, e^{-\frac{E}{kT}} \, dE$

Out[38]= $\dfrac{3 \text{ k T}}{2}$

The average kinetic energy from the Maxwell-Boltzmann distribution is seen to be in agreement with the definition of temperature (9.2.1).

9.3 QUANTUM DISTRIBUTIONS

The number density of particles with energy E may be written as

$$n(E) = \rho(E) f_{MB},$$

where $\rho(E)$ is a function called the density of states (discussed further in 9.3.4) that gives the number of states per energy and f_{MB} is the probability factor that the state is occupied,

$$f_{MB} = \frac{1}{e^{E/kT}}.$$

It is exponentially rarer to be in a state of higher energy. This holds true at lower energies (longer wavelengths), only if the particles are distinguishable, *i.e.*, the average distance between particles is larger than their wavelength as calculated for an ideal gas in 9.2.3.

In the quantum world where the wave properties are important, the particles are no longer distinguishable by their position and there are two classes: particles like the electron that obey the exclusion principle, and particles like the photon that do not.

9.3.1 Bose-Einstein Distribution

The Bose-Einstein (BE) distribution function applies to particles with integer spin, namely the photon. Its properties are derived from the fact that, unlike electrons, there is no exclusion principle preventing a large number of photons from occupying the same state. The BE distribution is subtly different from the MB distribution at low energies ($\frac{E}{kT} \ll 1$) where it becomes much larger. It is written

$$f_{BE} = \frac{1}{e^{E/kT} - 1}.$$

The main example of this is thermal radiation where the average energy per quantized oscillator was calculated (2.2) to be

$$\langle E_{osc} \rangle = \frac{E_p}{e^{E_p/kT} - 1},$$

where $E_p = hc/\lambda$ is the photon energy. The -1 in the denominator is crucial for understanding the distribution of photons at low energy.

9.3.2 Fermi-Dirac Distribution

The Fermi-Dirac (FD) distribution applies to particles with half-integer spin, namely the electron. No two electrons are allowed to be in the same state. The FD distribution is subtly different from the MB distribution at low energies, and is written

$$f_{FD} = \frac{1}{Ae^{E/kT} + 1}$$

where A is a parameter that is temperature dependent and is conventionally written as

$$A = e^{-E_F/kT}.$$

The constant E_F is called the Fermi energy, and it denotes the boundary between filled and open states at low temperature ($E_F \gg kT$). This FD distribution,

$$f_{FD} = \frac{1}{e^{(E-E_F)/kT} + 1},$$

gives approximately 1 when $E < E_F$ and 0 when $E > E_F$ (Fig. 9.8).

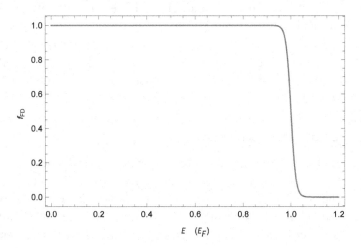

Figure 9.8 Fermi-Dirac distribution function. The energy is plotted in units of E_F

9.3.3 Comparison of the Distribution Functions

Figure 9.9 shows a comparison of the distribution functions at low energies. The FD distribution is unity (one electron per state) up to energy E_F (not shown), after which it drops rapidly to zero (see Fig. 9.8). The BE distribution goes to ∞ as the energy goes to 0.

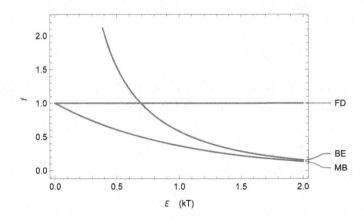

Figure 9.9 Comparison of the distribution functions at low energy. The FD distribution drops to zero quickly above the Fermi energy and the BE and MB distributions go to zero exponentially at large energy.

9.3.4 Density of States

The distribution functions give occupancy probability but do not give the number of states which is energy dependent. For that we need another function called the density of states. For an ideal gas, the density of states is proportional to the number density and $\sqrt{E}$. Using the result of 9.2.6, the density of states for the ideal gas is

$$\rho(E) = \frac{N}{V} \frac{2}{\sqrt{\pi}} (kT)^{-3/2} \sqrt{E}.$$

For distributions of electrons or photons, one can get the density of states using the concept of phase space, which is counting the number of waves that can fit inside a given volume in the same way as done in 2.1 with λ replaced by the de Broglie wavelength h/p. The number of states per volume is

$$n_{\mathrm{s}}(p) = g \frac{\frac{4}{3}\pi p^3}{h^3},$$

where the factor g accounts for the degeneracy of the states due to spin. The density of states is

$$\rho(E) = \frac{dn_{\mathrm{s}}}{dE} = \frac{dn_{\mathrm{s}}}{dp}\frac{dp}{dE} = g\frac{4\pi p^2}{h^3}\frac{d(\sqrt{E^2 - (mc^2)^2}/c)}{dE}.$$

Taking the derivative

$$\rho(E) = g\frac{4\pi p^2}{h^3}\frac{E}{pc^2} = g\frac{4\pi p^2}{h^3 v},$$

since the speed is given by $v = pc/E$. Thus, the density of states is proportional to the momentum squared divided by the speed.

9.3.5 Photon Gas

Applying this to photons, one gets

$$g = 2$$

for 2 possible polarizations, and

$$\frac{p^2}{v} = \frac{E^2}{c^3}.$$

The photon distribution becomes

$$n(E) = \rho(E)f_{BE} = \frac{8\pi E^2}{(hc)^3}\frac{1}{e^{E/kT} - 1}.$$

This is the number of photons per volume per energy that corresponds to thermal radiation. The energy per volume is E times this.

The energy density (u) of thermal radiation is (2.1.3)

$$u = \frac{4}{c}\sigma T^4.$$

The average photon energy is the ratio of total photon energy per volume divided by the number of photons per volume. The number per volume is obtained by integrating $n(E)$,

$$\langle E \rangle = \frac{\frac{4}{c}\sigma T^4}{\int_0^\infty n(E)dE}.$$

Example 9.27 Calculate the average energy of thermal photons as a function of T.

In[39]:= $Assumptions = {k > 0, T > 0};

$$N\left[\frac{4}{c}\frac{2\pi^5 k^4}{15 h^3 c^2}T^4 \Big/ \text{Integrate}\left[\frac{8\pi}{(hc)^3}\frac{E^2}{e^{E/(kT)} - 1}, \{E, 0, \infty\}\right], 3\right]$$

Out[39]= 2.70 k T

9.3.6 Electron Gas

For a nonrelativistic electron,

$$g = 2$$

for two spin states, and

$$\rho(E) = g\frac{4\pi p^2}{h^3 v} = 2\frac{4\pi}{h^3}\frac{2mE}{\sqrt{2E/m}} = \frac{4\pi(2m)^{3/2}}{h^3}\sqrt{E}.$$

The electron distribution becomes

$$n(E) = \frac{4\pi(2m)^{3/2}}{h^3}\sqrt{E}\frac{1}{e^{(E-E_F)/kT}+1}.$$

An expression for the Fermi energy may be obtained by integrating $n(E)$ to get the number density (N/V).

$$\frac{N}{V} = \int_0^\infty n(E)dE = \int_0^\infty \frac{4\pi(2m)^{3/2}}{h^3}\sqrt{E}\frac{1}{e^{(E-E_F)/kT}+1}dE.$$

The integration may be done at $T = 0$ because the Fermi energy does not depend on temperature as long as $E_F \gg kT$.

$$\frac{N}{V} = \frac{4\pi(2m)^{3/2}}{h^3}\int_0^{E_F}\sqrt{E}dE = \frac{4\pi(2m)^{3/2}}{h^3}\frac{2}{3}E_F^{3/2}.$$

Solving for the Fermi energy,

$$E_F = \frac{h^2}{8m}\left(\frac{3}{\pi}\right)^{2/3}\left(\frac{N}{V}\right)^{2/3}.$$

For a metal with one conduction electron, one may estimate

$$\frac{N}{V} \approx \frac{1}{(0.3 \text{ nm})^3}.$$

Example 9.28 Estimate the Fermi energy for a metal with one conduction electron per atom.

In[39]:= **m = [electron** PARTICLE **] [mass];**

$$\text{NumberForm}\left[\text{UnitConvert}\left[\frac{h^2}{8 m}\left(\frac{3}{\pi}\right)^{2/3}\left(\frac{1}{(0.3 \text{ nm})^3}\right)^{2/3}, \text{ eV}\right], 2\right]$$

Out[40]//NumberForm=

    4.1 eV

This is the correct order of magnitude. The Fermi energy for metals is a few eV. To calculate the Fermi energy more accurately, one may use the known density of the metal to get N/V.

Example 9.29 Calculate the Fermi energy of copper which has one conduction electron per atom.

In[41]:= $\mathbf{n} = \dfrac{\boxed{\textbf{copper}~\text{ELEMENT}}~\left[\boxed{\textit{mass density}}\right]}{\boxed{\textbf{copper}~\text{ELEMENT}}~\left[\boxed{\textit{atomic mass}}\right]}$;

$\mathbf{NumberForm}\left[\mathbf{UnitConvert}\left[\dfrac{\mathbf{h}^{2}}{\mathbf{8~m}}\left(\dfrac{3}{\pi}\right)^{2/3}(\mathbf{n})^{2/3},~\mathbf{eV}\right],~2\right]$

Out[42]//NumberForm=

7.0 eV

Example 9.30 Calculate the average energy of a conduction electron.

In[43]:= $\dfrac{\int_{0}^{E_F} E~\sqrt{E}~dE}{\int_{0}^{E_F}\sqrt{E}~dE}$

Out[43]=

$\dfrac{3~E_F}{5}$

9.3.7 Superfluid Helium

Helium is a boson with spin zero. Therefore, it obeys BE statistics. The density of states is similar to nonrelativistic electrons, except there is no factor of 2 for spin states, giving

$$n(E) = \frac{2\pi(2m)^{3/2}}{h^3}\sqrt{E}\frac{1}{e^{E/kT}-1}.$$

The number per volume is

$$\frac{N}{V} = \frac{2\pi(2m)^{3/2}}{h^3}\int_{0}^{\infty}\sqrt{E}\frac{1}{e^{E/kT}-1}dE.$$

Making a change of variables, $x = E/kT$,

$$\frac{N}{V} = \frac{2\pi(2mkT)^{3/2}}{h^3}\int_{0}^{\infty}\frac{\sqrt{x}}{e^{x}-1}dx.$$

Example 9.31 Evaluate the integral above.

In[44]:= $\texttt{NIntegrate}\left[\dfrac{\sqrt{x}}{e^x - 1}, \{x, 0, \infty\}\right]$

Out[44]=

2.31516

Solving for T,

$$T = \frac{h^2}{2mk}\left(\frac{(N/V)}{2\pi\,2.315}\right)^{2/3}.$$

which is the temperature at which liquid helium would obey BE statistics, *i.e.*, become a *superfluid*. This process is called a Bose-Einstein condensation.

Example 9.32 Evaluate the temperature at which He becomes a superfluid. Take the mass density at low temperature to be 140 kg/m³.

In[46]:= $\texttt{m} = \boxed{\texttt{helium}\ \text{ELEMENT}}\ \boxed{\texttt{atomic mass}}$;

$\texttt{n} = \dfrac{140\ \texttt{kg/m}^3}{\texttt{m}}$;

$\texttt{NumberForm}\left[\texttt{UnitConvert}\left[\dfrac{\texttt{h}^2}{2\ \texttt{m}\ \texttt{k}}\left(\dfrac{\texttt{n}}{2\pi\,(2.315)}\right)^{2/3}\right],\ 1\right]$

Out[48]//NumberForm=

3. K

The answer above is very close to the true value of 2.17 K. To get a more precise answer, one would need knowledge of a constant A which needs to be added to the BE probability (compare to the FD case where the Fermi energy was introduced),

$$f_{\mathrm{BE}} = \frac{1}{Ae^{E/kT} - 1}.$$

which has been set to 1 as in the case of the thermal distribution where it gave the correct answer.

The next candidate, after helium, to be a superfluid might be neon, but its melting point is too high , *i.e.*, it is a solid at the temperature it would need to reach to condense to a superfluid.

Example 9.33 Evaluate the temperature at which Ne would need to reach to become a superfluid and compare it to the melting point. Take the mass density at low temperature to be 1200 kg/m³.

In[49]:= **m** = [**neon** ELEMENT] [*atomic mass*] ;

$$n = \dfrac{1200 \; \dfrac{kg}{m^3}}{m} \; ;$$

$$\left\{ \text{NumberForm} \left[\text{UnitConvert} \left[\dfrac{h^2}{2\,m\,k} \left(\dfrac{n}{2\,\pi\,(2.315)} \right)^{2/3} \right], 1 \right] \right.$$

$$\left. , \; \text{UnitConvert} \left[[\text{neon} \; \text{ELEMENT}] [\text{melting point}], \; K \right] \right\}$$

Out[51]=

$$\{ 0.9 \, K, \; 24.56 \, K \}$$

CHAPTER 10

Astrophysics

10.1 THE SUN

10.1.1 Proton Cycle

The primary mechanism by which energy is produced in the sun is by the proton cycle. This is a multistep process in which the first step is the fusion of 2 protons. Two protons cannot make a stable bound state because of their electrical repulsion, but the bound state of a proton and a neutron, a deuteron (d), is stable. The reaction is

$$p + p \rightarrow d + e^+ + \nu_e.$$

The neutrino (ν_e) is produced to conserve lepton number. The kinetic energy released is the sum of the mass energies on the left minus that on the right. the neutrino is essentially massless. This may be followed by

$$p + d \rightarrow He^3 + \gamma,$$

and

$$He^3 + He^3 \rightarrow p + p + \alpha,$$

The net reaction is

$$4p \rightarrow \alpha + 2e^+ + 2\nu_e + 3\gamma.$$

There are other variances that have the same net reaction. The positron annihilates with an electron to give its mass energy back,

$$e^+ + e^- \rightarrow \gamma + \gamma,$$

DOI: 10.1201/9781003395515-10

Example 10.1 Calculate the amount of energy released in the proton cycle.

In[1]:= $N\Big[UnitConvert\Big[\Big(4\;\boxed{proton\;\text{PARTICLE}}\Big)\Big[\boxed{mass}\Big]$
$-\;m_\alpha\;+2\;\boxed{electron\;\text{PARTICLE}}\Big[\boxed{mass}\Big]\Big)\;c^2,\;MeV\Big],\;3\Big]$

Out[1]= 26.7 MeV

10.1.2 Distance to the Sun

The distance from the earth to the sun is called 1 astronomical unit (au).

Example 10.2 Convert au to m.

In[2]:= $N\big[UnitConvert\big[\;au\;,\;m\big]\big]$

Out[2]= 1.49598×10^{11} m

The parsec (pc) is a length unit defined by the distance at which the mean radius of the earth's orbit subtends an angle of one second of arc.

Example 10.3 Convert pc to m.

In[3]:= $N\big[UnitConvert\big[\;pc\;,\;m\big],\;3\big]$

Out[3]= 3.09×10^{16} m

The Mpc is in common use in astrophysics.

10.1.3 Solar Constant

The amount of solar energy reaching the earth, the solar constant (f) is measured to be

$$f = 1.36 \times 10^3 \text{ W/m}^2.$$

The total power output if the sun (luminosity) is

$$L_s = 4\pi(1 \text{ au})^2 f.$$

Example 10.4 Calculate the solar luminosity.

In[4]:= $UnitConvert\Big[\;solar\;constants\;4\,\pi\,(1\;au\,)^2,\;W\Big]$

Out[4]= 3.83×10^{26} W

10.1.4 Temperature of Sun

We may use the solar luminosity together with the Stefan-Boltzmann law to calculate the surface (blackbody) temperature of the sun.

$$R = \sigma T^4 = \frac{L_s}{4\pi r_s^2},$$

where r_s is the radius of the sun.

Example 10.5 Calculate the temperature of the sun.

In[5]:= $\text{UnitConvert}\left[\left(\dfrac{\boxed{\text{Sun } \text{STAR}}\left[\boxed{luminosity}\right]}{4\pi\sigma\,\boxed{\text{Sun } \text{STAR}}\left[\boxed{average\ radius}\right]^2}\right)^{\frac{1}{4}}\right]$

Out[5]= 5772. K

10.1.5 Neutrino Flux from the Sun

The sun produces energy by fusing protons (see 10.1.1). Each proton cycle releases 27 MeV of energy and produces 2 neutrinos.

Example 10.6 Calculate the solar neutrino flux at the surface of the earth.

In[6]:= $\text{N}\left[\text{UnitConvert}\left[2\,\dfrac{\text{solar constants}}{27\ \text{MeV}}\right], 1\right]$

Out[6]= $6. \times 10^{14}$ per meter^2 per second

That is an enormous number of neutrinos, often pointed out to be about 10^{11} through a thumbnail every second!

10.2 MAGNITUDE SCALE FOR SKY OBJECTS

10.2.1 Apparent Magnitude

Apparent magnitude is a measure of how bright a star or other sky object appears. The brightness depends on the luminosity, distance, and how much light is absorbed by material between the earth and the object. The brightness is characterized by the apparent magnitude parameter m on a reverse logarithmic scale such that brighter objects have smaller numbers. Apparent magnitude is scaled such that 5 units of m correspond to a factor of 100 in brightness. The brightest stars in the night sky are approximately $m = 1$ and the weakest to the naked eye are about $m = 6$. By comparison, the sun has $m = -26.7$.

10.2.2   Absolute Magnitude

Example 10.7 Convert Mpc to light years.

In[7]:= $N\left[\text{UnitConvert}\left[\text{Mpc}, \text{ly}\right], 3\right]$

Out[7]= 3.26×10^6 ly

The absolute magnitude (M) is defined as the apparent magnitude an object would have if viewed from a distance of 10 pc. For the The equation that relates apparent magnitude m and absolute magnitude M of a star at a distance d is called the distance modulus:

$$M - m = -5 \log_{10} \frac{d}{10\,\text{pc}}.$$

Example 10.8 Calculate the absolute magnitude of the sun.

In[8]:= $m = -26.74; d = \boxed{\textbf{Earth}\ \text{PLANET}}\boxed{\textit{distance from Sun}};$

$\text{DecimalForm}\left[m - 5\,\text{Log}\left[10, \dfrac{d}{10\,\text{pc}}\right], 3\right]$

Out[9]//DecimalForm=
     4.8

10.3   THE MILKY WAY

The distance of the sun from the center of the galaxy is measured by motion of the stars. The answer is about 8 kpc. The rotation speed of the sun about the galactic center is 2.3×10^5 m/s. The age of the sun, estimated by radioactive dating of the oldest meteorites, is about 4.6 By.

Example 10.9 Calculate the number of rotations the sun has made about the Milky Way center.

In[10]:= $v = 2.5 \times 10^5$ m/s ; $t = 4.6 \times 10^9$ yr ; R = 8 kpc ;

$\text{NumberForm}\left[\dfrac{v\,t}{2\,\pi\,R}, 2\right]$

Out[11]//NumberForm=
     23.

Newton's law for uniform circular motion gives

$$\frac{GM_g m_s}{R^2} = \frac{m_s v^2}{R}.$$

This gives the galactic mass (M_G) in terms of the solar mass (m_s) to be

$$\frac{M_G}{m_s} = \frac{R v^2}{G m_s}.$$

Example 10.10 Calculate the galactic mass in terms of the solar mass.

In[12]:= $\texttt{NumberForm}\left[\texttt{UnitConvert}\left[\dfrac{R\, v^2}{G\,\boxed{\textbf{Sun}\ \text{STAR}}\left[\boxed{mass}\right]}\right], 1\right]$

Out[12]//NumberForm=
$$1. \times 10^{11}$$

10.4   WHITE DWARF

A star that has consumed all its fusion energy may collapse to become a white dwarf which is stable against further gravitational collapse by electron pressure caused by the Pauli exclusion principle. A white dwarf is very dense having a solar mass in an earth size.

Example 10.11 Estimate the density of a white dwarf is solar masses.

In[13]:= $\texttt{N}\left[\texttt{UnitConvert}\left[\dfrac{\boxed{\textbf{Sun}\ \text{STAR}}\left[\boxed{mass}\right]}{\frac{4}{3}\pi\left(\boxed{\textbf{Earth}\ \text{PLANET}}\left[\boxed{average\ radius}\right]\right)^3}\right], 1\right]$

Out[13]=
$$2. \times 10^9\ \text{kg}/\text{m}^3$$

There is a maximum mass to the white dwarf beyond which it will collapse further to form a neutron star or black hole. This maximum mass is called the Chandrasekhar limit, the order-of-magnitude of which is given by

$$M \sim \frac{1}{m_H^2}\left(\frac{\hbar c}{G}\right)^{3/2}.$$

Example 10.12 Estimate the maximum size of a white dwarf.

In[14]:= **ScientificForm**$\Big[$

$\quad$ **UnitConvert** $\Big[\Big(\dfrac{1}{0.94 \text{ GeV} / c^2} \Big)^2 \Big(\dfrac{\hbar c}{G} \Big)^{3/2} \Big/$ $\boxed{\text{Sun } \text{STAR}} \big[\boxed{\textit{mass}} \big] \Big]$,

$\quad$ 2$\Big]$

Out[14]//ScientificForm=

$\quad$ 1.8

A more precise calculation gives about 1.4 solar masses for the maximum white dwarf mass.

10.5 NEUTRON STAR

A neutron star is formed from the gravitational collapse of a massive star. Further gravitational collapse is halted by the strong force when the neutron star density reaches that of nuclear density.

10.5.1 Density

Example 10.13 Estimate the neutron star density in kg/m^3 and (short) tons per tsp.

In[15]:= $\Big\{$ **ScientificForm** $\Big[$ **UnitConvert** $\Big[\rho = \dfrac{0.94 \text{ GeV} / c^2}{\frac{4}{3} \pi \big(1 \text{ fm} \big)^3} \Big], 1 \Big]$,

$\quad$ **ScientificForm** $\Big[$ **UnitConvert** $\Big[\rho, \dfrac{\text{sh tn}}{\text{tsp}} \Big], 1 \Big] \Big\}$

Out[15]=

$\quad \Big\{ 4. \times 10^{17} \text{ kg/m}^3 , 2. \times 10^9 \text{ sh tn/tsp} \Big\}$

That's a whopping 2 billion tons per teaspoon!

10.5.2 Fermi Energy

Example 10.14 Estimate the Fermi energy (see 9.3.6) for a neutron star.

In[16]:= **m** = [**neutron** PARTICLE] [*mass*];

$$\text{NumberForm}\left[\text{UnitConvert}\left[\frac{h^2}{8\,m}\left(\frac{3}{\pi}\right)^{2/3}\left(\frac{1}{\frac{4}{3}\,\pi\,(1\,\text{fm})^3}\right)^{2/3},\,\text{MeV}\right],2\right]$$

Out[16]//NumberForm=

   76. MeV

10.5.3 Binding Energy

The gravitational potential energy for a differential mass dm separated a distance r from. a mass m is

$$dU = \frac{Gm\,dm}{r}.$$

The mass is related to the density by

$$m = \frac{4}{3}\pi r^3 \rho,$$

so

$$dU = \frac{G}{\left(\frac{3}{4\pi\rho}\right)^{1/3}}m^{2/3}\,dm.$$

Assuming constant density which can be written in terms of the neutron star mass M and radius R ($\rho = \frac{M}{\frac{4}{3}\pi R^3}$), the binding energy is

$$E_b = \frac{G}{\left(\frac{3}{4\pi\rho}\right)^{1/3}}\int_0^M m^{2/3}\,dm = \frac{G}{\left(\frac{3}{4\pi\rho}\right)^{1/3}}\left(\frac{3}{5}\right)M^{5/3} = \frac{3GM^2}{5R}.$$

Example 10.15 Calculate the binding energy of a neutron star with 1.4 solar masses and a radius of 10 km.

In[17]:= **M** = 1.4 [**Sun** STAR] [*mass*]; **R** = 10 km ;

$$\text{ScientificForm}\left[\text{UnitConvert}\left[\frac{3\,G\,M^2}{5\,R},\,J\right],1\right]$$

Out[18]//ScientificForm=

   $3. \times 10^{46}$ J

10.6   BLACK HOLES

10.6.1   Schwarzschild Radius

For a mass M, the distance beyond which no particle, including photons, can escape. may be calculated in general relativity and is called the Schwarzschild radius (r_s).

$$r_s = \frac{2GM}{c^2}.$$

This gives the formula for the size of a black hole *vs.* mass.

Example 10.16 Calculate the Schwarzschild radius of the sun.

In[19]:= $\text{N}\left[\text{UnitConvert}\left[\dfrac{2\ \text{G}\ \boxed{\textbf{Sun}\ \text{STAR}}\ \boxed{\textit{mass}}}{\text{c}^{\,2}}\right],\,2\right]$

Out[19]=

3.0×10^3 m

10.6.2   Hawking Radiation

A black hole can radiate when pair production near the event horizon allows one particle to escape with the other falling into the black hole. The energy for the process comes from the gravitational field of the black hole. The resulting radiation follows the blackbody formula with temperature

$$T = \frac{\hbar c^3}{8\pi k G M}.$$

Example 10.17 Find the blackbody temperature of a black hole of mass 3 times that of the sun

In[20]:= $\text{N}\left[\text{UnitConvert}\left[\dfrac{\hbar\ \text{c}^{\,3}}{24\ \pi\ \text{k}\ \text{G}\ \boxed{\textbf{Sun}\ \text{STAR}}\ \boxed{\textit{mass}}}\right],\,2\right]$

Out[20]=

2.1×10^{-8} K

The radiated power per area is given by the Stefan-Boltzmann law, where the area comes from the Schwarzschild radius. The equation is

$$\frac{d(Mc^2)}{dt} = -4\pi\sigma r_s^2 \left(\frac{\hbar c^3}{8\pi kGM}\right)^4 = -\frac{c^8\hbar^4\sigma}{256\pi^3 k^4 G^2 M^2},$$

or

$$\int_{M_0}^0 M^2 dM = -\frac{c^8\hbar^4\sigma}{256\pi^3 k^4 G^2}\int_0^\tau dt,$$

where M_0 is the initial mass. The solution for the blackhole lifetime is

$$\tau = \frac{256\pi^3 k^4 G^2 M_0^3}{3c^6\hbar^4\sigma}.$$

Example 10.18 Calculate the lifetime of a black hole having a mass of 3 solar masses.

In[21]:= $\mathbf{N}\left[\text{UnitConvert}\left[\frac{256\ \pi^3\ \mathbf{k}^4\ \mathbf{G}^2\ \left(3\ \boxed{\text{Sun STAR}}\left[\boxed{\text{mass}}\right]\right)^3}{3\ c^6\ \hbar^4\ \sigma}, \text{yr}\right], 2\right]$

Out[21]=

5.7×10^{68} yr

10.7   THE DARK NIGHT SKY

Suppose stars with radius (R) and luminosity of the sun were distributed randomly with number density n in an infinite universe. The mean free path (see 3.7.4) of light before being absorbed and reradiated by a star is

$$d = \frac{1}{n\pi R^2}.$$

Example 10.19 Calculate the mean free path of starlight in an infinite universe. Take the visible mass density to be 3×10^{-28} kg/m^3.

In[22]:= $\mathbf{n} = \dfrac{3\times10^{-28}\ \frac{\text{kg}}{\text{m}^3}}{\boxed{\text{Sun STAR}}\left[\boxed{\text{mass}}\right]}$; $\mathbf{R} = \boxed{\text{Sun STAR}}\left[\boxed{\text{average radius}}\right]$;

$\mathbf{n} = \mathbf{N}\left[\text{UnitConvert}\left[\dfrac{1}{\mathbf{n}\ \pi\ \mathbf{R}^2}\right], 1\right]$

Out[23]=

$4. \times 10^{39}$ m

It would take 10^{31} s for light to travel from a star at distance d. The night sky would be bright because in any direction out to far enough distance there would be star in the line of sight. However, the universe is evolving on a time scale that is much faster than 10^{31} s and the night sky is dark. In other words there are no stars to be seen at such a distance.

10.8   HUBBLE'S LAW

Hubbles law gives the relationship between velocity, as determined by observed cosmological redshift, and distance

$$v = H_0 r,$$

where H_0 is called the Hubble constant.

10.8.1   Hubble Constant

The measured value of the Hubble constant is

$$H_0 = 73 \frac{\text{km/s}}{\text{Mpc}}.$$

The inverse of the Hubble constant is called the Hubble lifetime which is the expansion time of the universe, excluding the effect of gravitational pull.

Example 10.20 Calculate the Hubble lifetime.

In[24]:= $N\left[\text{UnitConvert}\left[\dfrac{1}{73 \text{ km}/ (\text{Mpc s})}, \text{ yr}\right], 2\right]$

Out[24]=

1.3×10^{10} yr

10.8.2   Cosmic Redshift

The relationship between redshift (z) and β (v/c) is

$$1 + z = \sqrt{\frac{1+\beta}{1-\beta}},$$

or

$$\beta = \frac{(1+z)^2 - 1}{(1+z)^2 + 1}.$$

Example 10.21 Calculate the distance to a galaxy which has a redshift of 1.

In[25]:= $z = 1; \; \beta = \dfrac{(1+z)^2 - 1}{(1+z)^2 + 1};$

$N\left[\text{UnitConvert}\left[\dfrac{c\,\beta}{73\ \text{km}/\,(\text{Mpc s})},\ \text{Mpc}\right],\ 2\right]$

Out[26]= 2.5×10^3 Mpc

10.9 COSMIC BACKGROUND RADIATION

The cosmic background radiation (CMB) is one of the cornerstones of cosmology. The CMB is left over from the early universe when radiation was in equilibrium with a hot, dense plasma of electrons, protons, and neutrons. When the plasma cooled, the first hydrogen and helium atoms were formed and the radiation decoupled from the charges that were no longer free. As the universe expanded, the CMB cooled to it present temperature of about 2.73 K.

The CMB spectrum fits that of a blackbody, and has been measured very accurately. The CMB temperature together with the Stefan-Boltzmann law gives the energy density (u) to be

$$u = \frac{4}{c}\sigma T^4.$$

Example 10.22 Calculate the energy density of cosmic photons.

In[26]:= $T = 2.726\ \text{K}\;;\ \text{NumberForm}\left[\text{UnitConvert}\left[\dfrac{4}{c}\,\sigma\,T^4,\ \dfrac{\text{MeV}}{\text{m}^3}\right],\ 2\right]$

Out[26]//NumberForm=
$0.26\ \text{MeV}/\text{m}^3$

In 2.2.5 the photon flux from blackbody radiation was calculated. The photon number density (n) in a cavity (in this case the universe) is given by

$$\frac{dn}{dE} = \frac{1}{E}\frac{du}{dE},$$

with

$$\frac{du}{dE} = \left(\frac{4}{c}\right)\left[\frac{2\pi E^3}{h^3 c^2 (e^{\frac{E}{kT}} - 1)}\right].$$

One integrates over all energies to get n. With the change invariables, $x = \frac{E}{kT}$,

$$n = 8\pi \left(\frac{kT}{hc}\right)^3 \int_0^\infty \frac{x^2}{e^x - 1}.$$

Example 10.23 Calculate the CMB photon number density.

In[27]:= $\texttt{NumberForm}\left[\texttt{UnitConvert}\left[\frac{4}{c} \; \frac{2\,\pi}{h^3 \, c^2} \; (k \; T)^3 \int_0^\infty \frac{x^2}{e^x - 1} \, dx\right], 3\right]$

Out[27]//NumberForm=

    4.11×10^8 per meter^3

10.10 COSMIC NEUTRINO BACKGROUND

At a temperature corresponding to kT = 2.5 MeV, neutrinos would decouple from matter. After expansion from that extremely early time to now, the universe is predicted to be left with a cosmic neutrino background (CNB) at a temperature of 1.95 K. The CNB is a black body distribution analogous to the CMB and the same technique (10.9) can be used to calculate their numbers. The CNB has not been observed due to the low neutrino interaction probability.

Example 10.24 Calculate the energy density of cosmic neutrinos.

In[28]:= $\texttt{T = 1.95 K ; NumberForm}\left[\texttt{UnitConvert}\left[\frac{4}{c} \; \sigma \; T^4, \; \frac{MeV}{m^3}\right], 2\right]$

Out[28]//NumberForm=

    0.068 MeV/m^3

Example 10.25 Calculate the CNB photon number density.

In[29]:= $\texttt{NumberForm}\left[\texttt{UnitConvert}\left[\frac{4}{c} \; \frac{2\,\pi}{h^3 \, c^2} \; (k \; T)^3 \int_0^\infty \frac{x^2}{e^x - 1} \, dx\right], 3\right]$

Out[29]//NumberForm=

    1.5×10^8 per meter^3

10.11 CRITICAL MASS DENSITY

The critical mass density (ρ_c) is defined to be the mass density of the universe beyond which expansion cannot continue forever. It is given by the condition on escape velocity

$$\frac{1}{2}mv^2 = \frac{GMm}{R},$$

Hubble's law

$$v = H_0 R,$$

and definition of mass density

$$\rho_c = \frac{M}{\frac{4}{3}\pi R^3}.$$

The result is

$$\rho_c = \frac{3H_0^2}{8\pi G}.$$

Example 10.26 Solve for the critical density.

```
In[30]:= ClearAll["Global`*"];

       Solve[1/2 m v^2 == (G M m)/R && v == Hθ R && ρc == M/(4/3 π R^3), {ρc, M, R}]

Out[31]= {{ρc → (3 Hθ^2)/(8 G π), M → v^3/(2 G Hθ), R → v/Hθ}}
```

Example 10.27 Calculate the critical mass density.

```
In[31]:= UnitConvert[(3 Hθ^2)/(8 π G), GeV/(c^2 m^3)]

Out[31]=
       5. GeV/(m^3 c^2)
```

This is about 5 hydrogen atoms per m^3.

10.12 PLANCK MASS

The Planck scale is defined by the energy that gives a dimensionless strength (analogous to electromagnetic α) of unity for gravity. It is an energy scale where the physics is unknown. The condition is

$$\frac{\hbar c}{Gm_{P}c^{2}} = 1.$$

Example 10.28 Calculate the Planck mass.

In[32]:= m_p = N[UnitConvert[$\sqrt{\dfrac{\hbar\ c}{G}}$, $\dfrac{GeV}{c^2}$], 2]

Out[32]=

1.2×10^{19} GeV/c^2

10.12.1   Planck Length and Planck Time

Apart from a factor of 2π, the Planck length is the wavelength of a photon that has an energy of M_Pc^2.

Example 10.29 Calculate the Planck length.

In[33]:= N[UnitConvert[$\dfrac{\hbar\ c}{m_p\ c^2}$], 2]

Out[33]=

1.6×10^{-35} m

The Planck time is the time it would take a photon of energy M_Pc^2 to travel a distance equal to the Planck length.

Example 10.30 Calculate the Planck time.

In[34]:= N[UnitConvert[$\dfrac{\hbar}{m_p\ c^2}$], 2]

Out[34]=

5.4×10^{-44} s

10.12.2 Relationship to Schwarzschild Radius

At the Planck mass, the Compton wavelength becomes

$$\lambda_c = \frac{hc}{M_P c^2} = \frac{hc}{c^2 \sqrt{\frac{hc}{2\pi G}}} = \frac{\sqrt{2\pi hcG}}{c^2},$$

and the Schwarzschild radius is

$$r_s = \frac{2GM_P}{c^2} = \frac{2G\sqrt{\frac{hc}{2\pi G}}}{c^2} = \frac{\sqrt{\frac{2}{\pi}hcG}}{c^2}.$$

Thus, they are the same order of magnitude.

Mathematica Starter

A.1 CELLS

Mathematica notebooks have 2 types of cells: "text" cells that can only display what is written and "input" cells that are executable. Input cells are executed by typing (simultaneously)

$$\boxed{\text{SHIFT}}\,\boxed{\text{RETURN}}$$

which generates the label In[ ]:= with the output going to another input cell with the label Out[ ]= (but it is still an input cell and can be executed).

Example A.1 Calculate 1+1.

```
In[1]:=   1 + 1
Out[1]=   2
```

A semicolon after a line of code means the code will still execute but the output will be suppressed. This is a useful feature for debugging code.

Example A.2 Set $x = 7$ and $y = 2$ and output $x + y$.

```
In[2]:=   x = 7; y = 2;
          x + y
Out[3]=   9
```

When a variable is set, its value may be used in other cells until cleared.

DOI: 10.1201/9781003395515-A

A.2 PALETTES

The menu has several extensive palettes that are useful in formatting the input. For example, there is a Writing Assistant that manages cells and fonts. This is useful for quick access to Greek letters. There is a Math Assistant that has templates for operations like division, raising to a power, summation, integration, *etc.* This makes it easy to enter something like a summation of squares into an input cell in a very clean format.

Example A.3 Sum the squares of integers from 0 to 10.

In[4]:= $\displaystyle\sum_{n=0}^{10} n^2$

Out[4]= 385

A.3 FUNCTIONS

Mathematica functions always begin with a capital letter. When typing in a cell, Mathematica will give autocomplete options for existing functions. Mousing over a function gives extensive documentation for the function's use with examples. The function Clear [ ] clears a variable. It produces no output.

Example A.4 Set $x = 1$ and clear x.

In[5]:= **x = 1**

Clear[x]

x

Out[5]= 1

Out[7]= x

The function ClearAll[ ] is extremely useful to perform a global clear of everything.

Example A.5 Clear all variables.

In[8]:= **ClearAll["Global`*"]**

The function Simplify[ ] reduces the result algebraically.

Example A.6 Calculate $\sin^2 x + \cos^2 x$.

In[10]:= `Simplify[Cos[x]^2 + Sin[x]^2]`

Out[10]=

    1

It can also be written as //Simplify, placed after the code.

Example A.7 Simplify $\frac{x^2-4x+4}{x-2}$.

In[11]:=
$$\frac{x^2 - 4x + 4}{x - 2} \text{ // Simplify}$$

Out[11]=

    $-2 + x$

The function D[ ] gives the dervative.

Example A.8 Calculate $\frac{d\cos x}{dx}$.

In[12]:= `D[Cos[x], x]`

Out[12]=

    $-\text{Sin}[x]$

A user may define a function by placing an underscore after the argument, $f[x\_]$. This allows the function to be evaluated for any value of the argument.

Example A.9 Define the function $f(x) = x^2$ and evaluate it for $x = 2.5$.

In[13]:= `f[x_] = x^2; f[2.5]`

Out[13]=

    6.25

A.4 RESERVED NAMES

There are a few names that are reserved and may not be user defined. One of them is D which is reserved for differentiation. The reserved names always begin with a capital letters. Others include, E , I, and Pi, which stand for the exponential e , imaginary i, and π.

Example A.10 Calculate $e^{i\pi} + 1$.

In[14]:= `X = E^{I Pi} + 1`

Out[14]=

    0

A double equal sign makes a logical comparison.

Example A.11 Compare Pi to π .

In[15]:= **Pi == π**

Out[15]=

True

A.5 PHYSICAL CONSTANTS AND THEIR UNITS

You can get physical constants in Mathematica by typing simultaneously (in input cell)

CTRL +

and then typing into the natural language box that appears, for example, "speed of light":

Figure A.1 Typing "speed of light" into the natural language box.

Clicking outside the box gives you (hopefully) what you were looking for, displayed in standard physics notation.

Figure A.2 Successful procurement of the speed of light.

The physical constant c is stored as a "unit" and it appears in italics with a different shading so you can recognize the difference between a unit and a user-defined variable with the same name. The numerical value is displayed together with units using the function UnitConvert[ ]. The default units will be SI.

Example A.12 Get the numerical value of the speed of light.

In[18]:= **UnitConvert[c]**

Out[18]=

299 792 458 m / s

The units to be displayed may be specified. There are 2 ways to get a unit 1) typing into the natural language box, and 2) using the function Quantity[ ].

Example A.13 Get the unit miles per second.

In[19]:= **Quantity["MilesPerSecond"]**

Out[19]=

1 mi / s

Example A.14 Get the numerical value of the speed of light in miles per second.

In[20]:= $x = \text{UnitConvert}\left[c , \dfrac{mi}{s} \right]$

Out[20]=

$$\dfrac{18\,737\,028\,625}{100\,584} \; mi/s$$

The function N[ ] will calculate the numerical value to the specified number of significant figures. The value has been stored in the variable x and it remains so until cleared.

Example A.15 Get the numerical value of the previous calculation of the speed of light to 3 significant figures.

In[21]:= **N[x, 3]**

Out[21]=

1.86×10^5 mi / s

Example A.16 Get π to 50 figures.

In[22]:= **N[π, 50]**

Out[22]=

3.1415926535897932384626433832795028841971693993751

The functions NumberForm [ ], and ScientificForm[ ] can also be used to display a decimal answer with specified number of digits.

Other useful physical constants are similarly obtained by typing the following into the natural language box: elementary charge, epsilon_0, planck's constant, hbar, electron mass, proton mass, boltzmann constant, *etc.* Physical constants and their names are given in App. B.

Mathematica is extremely useful as a calculator because it will automatically check the units of a calculation and report errors.

Example A.17 Try to get the speed of light in kg.

In[23]:= $\mathtt{UnitConvert}\big[\, c \, , \, kg \, \big]$

Out[23]=

$\mathtt{\$Failed}$

A.6 INTEGRATION

Integration is performed with the function Integrate[ ].

A.6.1 Indefinite Integrals

Example A.18 Calculate $\int x^3 e^{-x} dx$.

In[24]:= $\mathtt{Clear[x]; Integrate\big[x^3\ e^{-x}, x\big]}$

Out[24]=

$e^{-x}\left(-6 - 6\,x - 3\,x^2 - x^3\right)$

Mathematica can convert your input cell into a "standard form" which looks very much like you would see it typed in a book. This is equivalent code that executes identically. The computer code has become human readable!

Example A.19 Calculate $\int x^3 e^{-x} dx$ with the input in standard form.

In[25]:= $\int x^3\ e^{-x}\ dx$

Out[25]=

$e^{-x}\left(-6 - 6\,x - 3\,x^2 - x^3\right)$

A.6.2 Definite Integrals

Example A.20 Calculate $\int_0^\infty x^3 e^{-x} dx$.

In[26]:= $\mathtt{Integrate\big[x^3\ e^{-x}, \{x, 0, \infty\}\big]}$

Out[26]=

6

Example A.21 Calculate $\int_0^\infty x^3 e^{-x} dx$ with the input in standard form.

In[27]:= $\int_0^\infty x^3\, e^{-x}\, dx$

Out[27]=
    6

A.6.3 Numerical Integration

Numerical integration is performed with the function NIntegrate[ ].

Example A.22 Numerically integrate $\int_0^\infty \frac{x^2}{e^x-1} dx$.

In[28]:= $\text{NIntegrate}\left[\dfrac{x^2}{e^x-1},\ \{x,\ 0,\ \infty\}\right]$

Out[28]=
    2.40411

A.6.4 Assumptions

Mathematica will not assume that a variable is real.

Example A.23 Calculate the average value of $e^{-\lambda x}$ for $0 < x < \infty$.

In[29]:= $\dfrac{\int_0^\infty x\, e^{-\lambda x}\, dx}{\int_0^\infty e^{-\lambda x}\, dx}$

Out[29]=
    $\dfrac{1}{\lambda}$ if $\text{Re}[\lambda] > 0$

Assumptions about variables can be made with the global command $As-sumption. The can also make the code run faster for involved calculations.

Example A.24 Calculate the average value of $e^{-\lambda x}$ for $0 < x < \infty$ with the assumption that $\lambda > 0$.

In[30]:= $\text{\$Assumptions} = \lambda > 0;\ \dfrac{\int_0^\infty x\, e^{-\lambda x}\, dx}{\int_0^\infty e^{-\lambda x}\, dx}$

Out[30]=
    $\dfrac{1}{\lambda}$

A.7 RESOURCE FUNCTIONS

Mathematica has a large library of resource functions. These are called with the function ResourceFunction[ ] with the resource name in quotes, for example ResourceFunction["HydrogenWavefunction"].

Example A.25 Get the $2p$, $m_\ell = 0$ wave function of hydrogen with Bohr radius a.

In[31]:= `ResourceFunction["HydrogenWavefunction"]`

Out[31]=

[■] HydrogenWavefunction ✦

In[32]:= [■] HydrogenWavefunction ✦ `[{2, 1, 0}, a, {r, θ, φ}]`

Out[32]=

$$\frac{\sqrt{\frac{1}{a^3}} \; e^{-\frac{r}{2a}} \; r \; \text{Cos}[\theta]}{4 \, a \, \sqrt{2\pi}}$$

A.8 SERIES EXPANSION

Example A.26 Get the first 4 terms (order x^3) of $\sin x$ expanded about $x = 0$.

In[33]:= `Series[Sin[x], {x, 0, 4}]`

Out[33]=

$$x - \frac{x^3}{6} + O[x]^5$$

Example A.27 Get the first 3 terms of the binomial expansion.

In[34]:= `Series[(1 + x)^n, {x, 0, 3}]`

Out[34]=

$$1 + n\,x + \frac{1}{2}\,(-1+n)\,n\,x^2 + \frac{1}{6}\,(-2+n)\,(-1+n)\,n\,x^3 + O[x]^4$$

A.9 SOLVING AN EQUATION

Example A.28 Solve $x^2 - 4x + 2 = 0$.

In[35]:= `y = Solve[x^2 - 4 x + 2 == 0, x]`

Out[35]=

$$\left\{\left\{x \to 2 - \sqrt{2}\right\}, \left\{x \to 2 + \sqrt{2}\right\}\right\}$$

Solve produces a list. It is useful sometimes to put the results from Solve into a variable.

Example A.29 Extract the results from Solve above for the solution to x stored in the variable y.

In[36]:= **x /. y[[1]]**

x /. y[[2]]

Out[36]=

$$2 - \sqrt{2}$$

Out[37]=

$$2 + \sqrt{2}$$

A.10 PLOTTING A FUNCTION

Plot[Sin[x], {x, 0, 6π}, PlotStyle→GrayLevel[.5],

AxesLabel → {"x", "sin(x)"}]

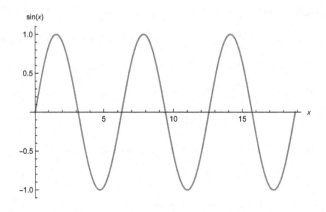

Figure A.3 Plot of $\sin x$ vs. x.

Physical Constants

Physical constants may be called in two ways. One way is to use the function Quantity[ ] with the argument equal to the name of the constant. A second way which makes a much cleaner look to the code is to use the defined symbol obtained from the natural language box as described in A.5. Executing Quantity[ ] produces the an output identical to that of the natural language box.

Example B.1 Compare the elementary charge as obtained from the Quantity[ ] function and the natural language box.

In[1]:= **Quantity["ElementaryCharge"]** == ⊟ *e* ⌊···⌋ ✓

Out[1]= **True**

The names of the fundamental constants used in this book with their symbols and values are shown in B.2 and derived combinations are shown in B.1.

Particle masses may be acquired with either the function Quantity[ ] or the natural language box as described in A.5.

Example B.2 Get the numerical value of the electron mass using the natural language box.

In[2]:= **ScientificForm**⌊**UnitConvert**⌊ electron PARTICLE ⌋⌊ *mass* ⌋⌋**, 4**⌋

Out[2]//ScientificForm=
    9.109×10^{-31} kg

Particle masses are displayed in B.3.

DOI: 10.1201/9781003395515-B

Table B.1 Mathematica names (symbol) and numerical values for physical constants.

Physical Constant	Value
ElementaryCharge (e)	1.60218×10^{-19} C
ElectricConstant (ε_0)	$8.85418781 \times 10^{-12}$ C/(m V)
MagneticConstant (μ_0)	$1.256637062 \times 10^{-6}$ m T/A
SpeedOfLight (c)	299 792 458 m/s
PlanckConstant (h)	4.13567×10^{-15} s eV
GravitationalConstant (G)	6.6743×10^{-11} m^3/(kg s^2)
BoltzmannConstant (k)	0.0000861733 eV/K
AvogadroNumber (N_0)	6.02214×10^{23}

Table B.2 Mathematica names (symbol) and numerical values for derived constants.

Derived Constant	Value
ReducedPlanckConstant ($\hbar$)	6.58212×10^{-16} s eV
FineStructureConstant (α)	0.007297352569
ElectronComptonWavelength(λ_e)	$2.42631024 \times 10^{-12}$ m
BohrRadius (a_0)	0.0529177211 nm
RydbergConstant (R_∞)	1.09737×10^7 per meter
BohrMagneton (μ_B)	0.0000578838 eV/T
StefanBoltzmannConstant (σ)	5.67037×10^{-8} W/(m^{2}K^4)

Table B.3 Mathematica names (symbol) and numerical values for particle masses.

Particle	Mass
ElectronMass (m_e)	0.510999 MeV/c^2
MuonMass (m_μ)	105.658 MeV/c^2
ProtonMass (m_p)	938.272 MeV/c^2
NeutronMass (m_n)	939.565 MeV/c^2
DeuteronMass (m_d)	1875.61 MeV/c^2
AlphaParticleMass (m_α)	3727.38 MeV/c^2

Table B.4 Common names (symbol) and numerical values for sky objects.

Object	Mass
mass of Moon (Moon [mass])	7.3459×10^{22} kg
mass of Earth (Earth [mass])	5.9722×10^{24} kg
solar mass (Sun [mass])	1.98844×10^{30} kg
neutron star mass ((1.1 to 2.1) $M_\odot$)	(2.18724×10^{30} to 4.17564×10^{30}) kg
Milky Way mass (Milky Way [mass])	3.06214×10^{42} kg
mass of universe (m_U])	1.51184×10^{53} kg

Index

Printed in the United States
by Baker & Taylor Publisher Services